INSECTS AND OTHER TERRESTRIAL ARTHROPODS:
BIOLOGY, CHEMISTRY AND BEHAVIOR

SPIDERS

MORPHOLOGY, BEHAVIOR AND GEOGRAPHIC DISTRIBUTION

Insects and Other Terrestrial Arthropods: Biology, Chemistry and Behavior

Additional books in this series can be found on Nova's website under the Series tab.

Additional e-books in this series can be found on Nova's website under the e-book tab.

Animal Science, Issues and Professions

Additional books in this series can be found on Nova's website under the Series tab.

Additional e-books in this series can be found on Nova's website under the e-book tab.

INSECTS AND OTHER TERRESTRIAL ARTHROPODS:
BIOLOGY, CHEMISTRY AND BEHAVIOR

SPIDERS

MORPHOLOGY, BEHAVIOR AND GEOGRAPHIC DISTRIBUTION

MARQ SANTERRE
EDITOR

NOVINKA

New York

NOTICE TO THE READER

LIBRARY OF CONGRESS CATALOGING-IN-PUBLICATION DATA

ISBN: 978-1-62618-517-3

Published by Nova Science Publishers, Inc. † New York

CONTENTS

PREFACE

This book presents current research in the study of the morphology, behavior and geographic distribution of spiders. Topics discussed include plasticity in spider webs and silk; spider webs as extended phenotypes; the structure and function of the lipoproteins of spiders; a study of the discriminate information on spider webs; and ant mimicry in spiders.

Chapter 1 – One of the most striking aspects of spider biology is web building and silk spinning. While webs and silks vary widely among species, web architecture and silk mechanics are also plastic at the individual level. This review will focus on plasticity in web architecture and silk mechanics. Plasticity in web architecture exists in both orb webs and cobwebs. Web building, including the presence of a stabilimentum, is influenced by several environmental factors, including foraging success, prey type, presence of predators and spatial constraints. Various models, such as optimal foraging, optimal performance and physiological limitations, have attempted to explain web architecture plasticity. The mechanics and stickiness of silk is also influenced by the environment, including prey type and nutritional content, spider condition and silk use. An interesting phenomenon is major ampullate silk's response to water, termed supercontraction, and characterized by shrinking and rubberlike behaviour. Silk mechanics may change through changes in silk biochemistry or changes in silk dope processing. As web architecture and silk mechanics combine to determine web behaviour, web and silk plasticity may have consequences for whole web performance by affecting web's ability to sustain impacts, retain prey and transmit prey vibrations.

Chapter 2 – An animal's structure is an expression of its genotype beyond the confines of the body, so is considered an extended phenotype. The fitness consequences of extended phenotypes are not well defined because extended phenotypes affect fitness via interactions with both the environment and the somatic phenotype and these interactions are not well understood. Here the author illustrate that web-building spiders make ideal models for testing hypotheses regarding the functional significance and adaptive benefits of extended phenotypes. A review of the literature reveals that spider webs, silks and additional components, e.g. decorations, decoys and barrier webs, represent expressions of (i) a multitude of behaviours including foraging and courtship, (ii) the spider's ontogeny, (ii) the sensory capacities of the spider and/or its predators and prey, and (iv) the phenotypic expression of any infecting parasitoids. The author herein assess the adaptive benefits of expressing webs as extended phenotypes and outline the feedback cycles that influence spider fitness. Niche construction is an under explored aspect of extended phenotypes. The author thus describe how it might play a role in enhancing spider fitness. The author will lastly outline some research applications of using the 'web as an extended phenotype' model, in investigations of: (i) invasive and conservation biology of spiders, (ii) spider phylogenetics, and (iii) using webs for biomonitoring, among others.

Chapter 3 – It is well known that the Class Arachnida is one of the arthropod groups with major diversity in the world. There are around 100 families belonging to the Order Araneae; over 42,000 species from different geographical areas have been described. In the last few years biochemical/physiological studies in some species of spiders: *Polybetes pythagoricus, Latrodectus mirabilis, Eurypelma californicum, Cupiennius salei, Nephila inaurata* among others have significantly increased. Many of them are based on the venom characterization, but only a few of them are related to basic molecules such as lipids, proteins or carbohydrates.

Lipids are transported from the tissue of origin to those tissues where they are either stored or metabolized. In an aqueous environment as the hemolymph they must be associated to proteins, forming lipoproteins which may be more complex generating the so-called glycolipoproteins.

Within the Class Arachnida, different lipoproteins were identified and characterized in the Orders Araneae, Acarina, Scorpionida and Solpugida. In only three species of spiders, *Polybetes pythagoricus* and *Latrodectus mirabilis* (Suborder Araneomorphae) and *Eurypelma californicum* (Suborder Mygalomorphae), lipoproteins of different density were purified and studied. *P. pythagoricus* belongs to the Sparassidae family, and it is only found in South America, while *L. mirabilis* belongs to the Theridiidae family and is distributed in the center and south of Argentina and Uruguay. *E. californicum* belongs to the Theraphosidae family that is present in the American continent, mainly in North America. Lipoproteins of high density ranging from 1.07 to 1.14 g/ml have been detected as the major circulating lipoproteins in these species. Concerning the lipid composition of Araneomorphae, the major lipids are phosphatidylcholine, free fatty acids and triacylglycerides (being this one the major energetic lipid). As to Myghalomorphae *E. californicum*, this lipoprotein of high density named lipophorin as in insects, mainly contains phospholipids and diacylglycerol (the major energetic lipid). Another high density lipoprotein named HDL2 is found in *P. pythagoricus* and *L. mirabilis*, while *P. pythagoricus* and *E. californicum* have a third lipoprotein of very high density (VHDL) in hemolymph. It is important to remark that the hemocyanin in *P. pythagoricus* and *L. mirabilis*, in addition to its classical role as respiratory pigment, plays functions as apolipoprotein. Lately, *P. pythagoricus* has been used as a model in order to further study hemolymph lipoproteins (HDL1, HDL2 and VHDL). For the first time, egg glycolipoproteins of high and very high density (LV1 and LV2) which are responsible for the embryo nutrition, were identified in class Arachnida. Phospholipids are the major structural lipids. Oleic acid is the major component in both lipoproteins. These particles are spheroidal and their diameter ranges between 20 nm in LV1 and 27 nm in LV2.

Recently, the authors have reported data on the secondary and tertiary structure of VHDL by using spectroscopy and sequencing methods.

Despite the fact that biochemical and physiological studies dealing with lipoproteins have increased, further studies are necessary to elucidate the dynamics of proteins and lipids in arachnids.

Chapter 4 – This work presents the demonstration of the biological hypothesis about the localization of the main information of the web spider, in order to apply it for identifying of spider species. The question is the analysis of the whole the web spider or only the center of the spider web, as the hypothesis says. After the authors' experiments with their database, applying a pre-processing stage, enhanced the spider web images and using different kinds of parameterization methods for classifying it with support vector machines; this work have demonstrated the hypothesis is true, and the most important information is located on the center of the spider web.

Chapter 5 – Mimicry is a natural phenomenon which never ceases to amaze, involving one organism taking on characteristics of another, to gain evolutionary advantage. These strategies based on "deceit" carry with them advantages to the mimic (or deceiver) either in the form of heightened protection from potential predators (as in the case of Batesian or Müllerian mimicry), or easier access to prey (aggressive mimicry). Mimics exploit different sensory channels, including visual, acoustic or chemical systems, by looking, sounding or "smelling" like a model organism, respectively. Certain characteristics of model organisms are favorable to mimics, especially when the mimicry is a form of protection from predators. Among insects, an ideal group of model organisms are the ants, as they are numerous, often aggressive and unpalatable for many other animals. Ant mimicry is therefore common in several invertebrate families. In spiders, ant mimicry occurs in at least 13 different families. Here, ant mimicry in spiders will be outlined, briefly describing the main findings to date of this vast and fascinating topic.

In: Spiders
Editor: Marq Santerre

ISBN: 978-1-62618-517-3

Chapter 1

PLASTICITY IN SPIDER WEBS AND SILK: AN OVERVIEW OF CURRENT EVIDENCE

***Cecilia Boutry*[1] *and Sean J. Blamires*[2]**
[1]Department of Integrative Biology, University of Guelph, Guelph ON, Canada
[2]Department of Life Science, Tunghai University, Taichung, Taiwan

ABSTRACT

One of the most striking aspects of spider biology is web building and silk spinning. While webs and silks vary widely among species, web architecture and silk mechanics are also plastic at the individual level. This review will focus on plasticity in web architecture and silk mechanics. Plasticity in web architecture exists in both orb webs and cobwebs. Web building, including the presence of a stabilimentum, is influenced by several environmental factors, including foraging success, prey type, presence of predators and spatial constraints. Various models, such as optimal foraging, optimal performance and physiological limitations, have attempted to explain web architecture plasticity. The mechanics and stickiness of silk is also influenced by the environment, including prey type and nutritional content, spider condition and silk use. An interesting phenomenon is major ampullate silk's response to water, termed supercontraction, and characterized by shrinking and rubberlike behaviour. Silk mechanics may change through changes in silk biochemistry or changes in silk dope processing. As web architecture and silk mechanics combine to determine web behaviour, web and silk plasticity may have consequences for whole web performance by affecting web's ability to sustain impacts, retain prey and transmit prey vibrations.

INTRODUCTION

One of the remarkable aspects of spider biology is that most spiders rely on an external trap, the web, for catching their prey. Web architecture is extremely diverse among the 43,000+ extant species of spiders (Platnick, 2000-2010). The most well-known web type is the planar orb web. However, some species build three-dimensional webs, including cobwebs, funnel webs and sheet webs. Others, such as tarantulas and wolf spiders, do not build webs at all.

Webs are built using silk. The number of silk types a single spider can produce varies among species, and can reach as much as seven in orb-weaving spiders (Tillinghast and Townley, 1994). Silks are used for many purposes by spiders, from wrapping egg sacs to holding the spider if it drops. Orb webs are typically made of major ampullate silk (MAS), an “all-purpose” silk whose properties are well studied, and flagelliform silk, a highly extensible silk coated with glue derived from the aggregate glands (Denny, 1976; Foelix, 1987; Work, 1976). Each silk type has a unique amino acid composition and mechanical behaviour (Andersen, 1970; Blackledge and Hayashi, 2006; Gosline, et al., 1999). Nonetheless, this behaviour may vary among species. For instance, orb-weaving spiders have significantly stronger MAS than haplogyne spiders (Swanson, et al., 2006). Web and silk properties, however, may also vary between individuals.

Phenotypic plasticity (some authors may alternatively use terms such as ‘flexibility’ or ‘variation’ to differentiate with ‘plasticity’ that may be defined as the physical effects of the environment on organismal forms) is the ability of an individual organism to change its phenotypic expression (e.g. physiological expression, behaviours and morphology) in response to changing environments (Miner, et al., 2005; Pigliucci, 2005; Pigliucci, et al., 2006). Plastic phenotypes may vary in response to predators (inducible defenses) or prey (inducible offenses) presence, or abiotic environmental variables (Gabriel, 2005; Kishida, et al., 2010; Miner, et al., 2005; Mougi, et al., 2011; Pigliucci, et al., 2006). Additionally, the plastic response, itself, may vary depending on the influence of the context, strength, frequency and reliability of initiator cues and the strength of positive or negative feedback mechanisms (Auld, et al., 2009; Gabriel, 2005; Miner, et al., 2005).

We will firstly discuss web plasticity and compare the plasticity of orb-webs, cobwebs and other web types. We will present some explanative models that have been used to rationalize spider web plasticity in a behavioural, physiological or ecological context. We will then focus on silk plasticity. We will describe the mechanisms of plasticity in MAS tensile behaviour, including supercontraction, and capture spiral stickiness. We will conclude the chapter by examining how web architecture and silk mechanics combine to affect spider web function.

1. Plasticity in Web Architecture

Spider webs are described as having plasticity at three levels: (i) the physical and chemical properties of the component silks, which we will consider in detail later in this article; (ii) the expression of individual architectural components, and (iii) the performance of the web as a whole (Blackledge, et al., 2011; Craig, 2003; Harmer, et al., 2011; Herberstein and Tso, 2011). Each level directly or indirectly influences each other, and each may vary in response to a multitude of environmental stimuli, including changes in: (i) levels of predation threat, (ii) prey type, availability or nutrient quality, and (iii) a multitude of abiotic factors, e.g. wind and temperature.

1.1. Plasticity in Web Architectural Components

To examine the plasticity in expression of the individual components of spider webs in detail we will first elaborate on what those components are in the different types of spider web.

1.1.1. Components of Orb Webs

The architectural components of two-dimensional orb webs for which individual plasticity is traditionally measured are the: (i) capture surface area, (ii) width of the spaces between the sticky spirals (mesh size), (iii) number of radii that traverse the spiral thread, and (iv) length or pattern of any decorations (stabilimenta) (Blamires, 2010; Herberstein, et al., 2000a; Liao, et al., 2009; Mayntz, et al., 2009; Schneider and Vollrath, 1998; Tso, et al., 2007) (Figure 1). Plasticity in these parameters is generally considered to be the mean amount of variation across some kind of treatment(s). Each component consists of a different combination of silks that are named based on the silk gland from which they are derived (Blackledge and Hayashi, 2006; Vollrath and Knight, 2001). The radial threads are composed of major (MAS) and minor ampullate (MiA) silks (Craig, 2003; Vollrath and Knight, 2001). The spiral silks on the other hand are derived from either dry pseudoflagelliform cribellate silk or flagelliform silk coated with sticky aggregate silks (Opell and Bond, 2001; Tarakanova and Buehler, 2012), and the decorations are principally formed of aciniform silks (Craig, 2003; Tso, 2004; Walter, et al., 2008). Egg sacs, uneaten prey and inert materials can also be used to form decorations (Blamires, et al., 2010b; Gan, et al., 2010; Herberstein, et al., 2000a; Manicom, et al., 2008; Tan and Li, 2009; Tseng and Tso, 2009).

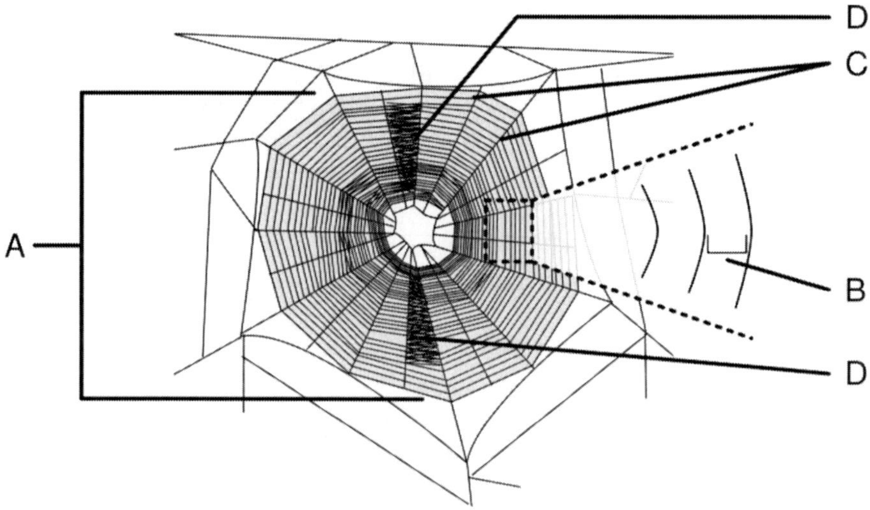

Figure 1. Diagram of a vertical two-dimensional orb web, showing the architectural components. capture surface area (A), space between the sticky spirals (i.e. mesh size) (B) , radii traversing the spiral thread (C), and decorations or stabilimenta (D).

Other orb web components that have been measured and may show inter-individual plasticity include the position of the hub and the resultant up-down asymmetry of the web (Coslovsky and Zschokke, 2009; Nakata, 2012; Nakata and Zschokke, 2010) and barrier webs and other add-ons (Baba and Miyashita, 2006; Blackledge and Wenzel, 2001a; Blackledge and Wenzel, 2001b; Higgins, 2006; Uhl, 2008). Aggregated webs may function similar to barrier webs, i.e. as protection for the spider or assistance in prey capture via "ricochet effects" (Blamires, et al., 2010b; Rao, 2009; Uetz, 1989), so may also be considered a kind of additional web component, but we will not specifically elaborate on these herein. Some orb webs, e.g. those built by spiders of the genera *Cyrtophora* and *Mecynogea.* include permanent three-dimensional webs with a horizontally placed orb (Blamires, et al., 2012b; Levi, 1997). We do not know of any studies examining web architectural plasticity in these spiders, so the following pertains only to two-dimensional orb webs.

1.1.2. Plasticity in Orb Web Components

The above-mentioned architectural components are readily identified and quantified directly from webs by counting threads, measuring widths and heights and calculating parameters using formulae (Blackledge and Gillespie, 2002; Herberstein and Tso, 2000; Sandoval, 1994; Tso, 1999). Factors thus far ascribed as influential over the expression of orb web components are many and include predators, supporting structural architecture, air temperature, wind, light levels, prey availability, prey types and nutrients (Blamires, et al., 2011; Harmer and Herberstein, 2010; Heiling and Herberstein, 2000; Hesselberg and Vollrath, 2006; Liao, et al., 2009; Sherman, 1994; Vollrath, et al., 1997).

Spiders that had recently fed will build a web with a smaller capture area than spiders who have not recently fed (Herberstein, et al., 2000c; Higgins and Buskirk, 1992; Venner, et al., 2000). There are optimal foraging explanations for this phenomenon (Blackledge, 2011; Herberstein, et al., 2000b; Herberstein, et al., 2000c; Venner and Casas, 2005; Venner, et al., 2000), which we will describe later. The capture surface area is often associated with a reduction in mesh size (Sandoval, 1994; Sherman, 1994; Tso, 1999; Watanabe, 2001), but mesh size may vary independently in certain circumstances, e.g. when a spider is starved or nutritiously depleted (Herberstein, et al., 2000c; Higgins and Rankin, 1999). If the capture surface area correlates with a reduction in mesh size, the total investment in capture thread is unchanged; i.e. smaller webs have more densely packed spirals (Sherman, 1994; Venner, et al., 2000). A change in spiral thread investment is implicit if mesh size remains unchanged but the capture surface area changes, or if the mesh size changes but the capture surface area remains unchanged (Blamires, 2010; Sensenig, et al., 2010; Sherman, 1994; Watanabe, 2001).

When a spider is forced to change its web capture surface area, for example because of changes in vegetative supporting structure, it must make a decision whether or not to concomitantly alter its mesh size or total spiral thread length (Blackledge and Eliason, 2007; Sandoval, 1994; Scharf, et al.,

2011; Tso, 1999). Generally, if the spider expects to feed on similar prey it will not change mesh size (Sandoval, 1994), so it will invest in a longer spiral thread to accommodate the web size change (Opell, 1999). Spiders are reluctant to alter the mesh size because mesh size is usually honed so that the web can catch the most abundant, or a preferred, type of prey in the vicinity (Blackledge and Zevenbergen, 2006; Blamires, et al., 2011; Herberstein and Heiling, 1998; Miyashita, 1997; Sandoval, 1994). There is now evidence, based on optimal foraging models, to suggest that orb web spiders gain fitness benefits if they build webs with mesh sizes capable of capturing the largest prey expected even if these prey are relatively rare (Blackledge, 2011; Blamires, et al., 2011; Sandoval, 1994; Venner and Casas, 2005). Regardless, some studies have shown that spiders will change the mesh size in their webs even if prey type or availability does not change, so factors other than prey, e.g. the presence of predators (Nakata, 2008), also influence the spiral thread investment decisions in orb web spiders. In contrast to spiral threads, more radial threads are generally added to an orb web when a spider is satiated compared to when it is starved (Blamires, et al., 2011; Mayntz, et al., 2009; Sherman, 1994; Tso, et al., 2007; Tso, 1999), or if the spider detects the presence of a potential predator (Nakata, 2007).

Spiders forced by spatial constraints to build smaller webs may concomitantly add more radii to their webs (Tarakanova and Buehler, 2012; Vollrath and Mohren, 1985). Nonetheless in accordance to ecological models it remains unclear why the extra investment is made.

Approximately 22 genera of diurnal orb-web spiders add conspicuous structures called decorations or stabilimenta (depending on the authority) to their webs using silk, egg sacs, uneaten prey and inert material (Blackledge, 1998; Bruce, 2006; Herberstein, et al., 2000a; Théry and Casas, 2009). Decorations have been studied for decades, but their function remains enigmatic (Blackledge, 1998; Herberstein, et al., 2000a; Walter and Elgar, 2012).

There are a number of hypotheses that attempt to explain why certain spiders use decorations, including prey-luring, predator avoidance, a "warning" or "advertising" system protecting the web against damage by passing animals, and a molting or thermoregulatory platform (see Bruce, 2006; Craig, 2003; Herberstein, et al., 2000a; Walter and Elgar, 2012 for comprehensive lists). Here we do not make any judgments on the validity of any of these hypotheses. Rather, we examine whether or not decoration forms vary in different environments and, if so, why they do.

Many spiders of the genus *Agiope*, construct decorations consisting of four zigzag silk bands arranged in a cruciate (x-shaped) form. The form of the decoration is highly variable within and among individuals; an individual spider may build a decoration with all four or 1, 2, 3 or 4 of the bands or no decoration at all on any day (Blamires, 2010; Blamires, et al., 2009; Blamires, et al., 2007; Craig, 1994b; Craig, et al., 2001; Rao, 2010; Seah and Li, 2002; Walter and Elgar, 2011).

Predator and/or prey learning of decoration designs, variations in the spider's nutritional intake, the spider's weight, human/animal induced web damage and the onset of the molting phase have all been touted as factors inducing variability in decoration designs in *Argiope* and other, e.g. *Octonoba*, spiders (Blackledge, 1998; Blackledge and Wenzel, 1999; Blamires, 2010; Blamires, et al., 2009; Blamires, et al., 2007; Craig, et al., 2001; Walter and Elgar, 2011; Watanabe, 1999). In the genus *Cyclosa* (who may build their decorations from plant debris, molted cuticle, and egg sacs or silk), decoration variability has been described as induced by variation in wind speed (Liao, et al., 2009) and in the presence of predators or upon detection of predator cues (Gan, et al., 2010; Nakata, 2009; Tseng and Tso, 2009).

1.1.3. Components of Cobwebs

Cobwebs are the three dimensional webs of spiders of the diverse family Theridiidae. We consider them separately to orb webs as their architecture is unique and their silks highly specialized (Barrantes and Weng, 2006; Blackledge, et al., 2005a; Blackledge, et al., 2005b; Eberhard, et al., 2008). Furthermore, there have been numerous studies explicitly examining cobweb architectural plasticity (Agnarsson and Coddington, 2006; Barghusen, et al., 1997; Blackledge and Zevenbergen, 2007; Jorger and Eberhard, 2006; Salomon, 2007; Zevenbergen, et al., 2008). These ascribe the three major variable components of a cobweb as: (i) the retreat, a small "pocket" consisting of supporting tangle threads and a non-sticky sheet where the spider can hide, (ii) supporting threads, or sheet, which form a large, relatively easily accessible, tangle.

Like the sheet, not every species makes a retreat (*Achaearanea* spp. do not make either), (iii) vertically descending gumfooted threads, and (iv) gumfoot sticky droplets deposited at the base of the gumfooted threads (Barrantes and Weng, 2006; Blackledge, et al., 2005a; Boutry and Blackledge, 2008, 2009; Eberhard, et al., 2008) (Figure 2).

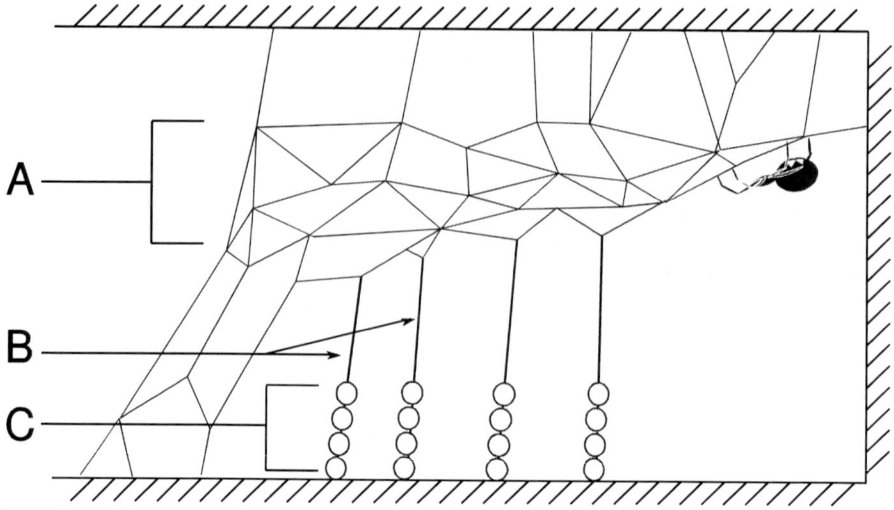

Reproduced from Blackledge and Zevenbergen, 2007.

Figure 2. Outline of a cobweb showing the retreat, the supporting tangle threads, the non-sticky sheet where the spider hides (A), and the vertically descending gumfooted threads (B) with sticky droplets at the base (C).

1.1.4. Plasticity in Cobweb Components

As with orb webs, spatial constraints (Jorger and Eberhard, 2006) hunger/satiation and type of food consumed influence cobweb architecture (Blackledge and Zevenbergen, 2007; Lubin and Henschel, 1996; Zevenbergen, et al., 2008). For instance, satiated *Latrodectus* spp. deposit less gumfooted threads, with each thread containing less sticky droplets, than do starved *Latrodectus* spp. (Blackledge and Zevenbergen, 2007; Lubin and Henschel, 1996). The sizes of the individual gumfoot droplets, however, do not seem to differ between satiated and hungry spiders (S.J. Blamires, unpublished data). Temperature and light also have been described to influence the architecture of cobwebs, most likely by directly affecting the spider's physiological ability to spin silk and build webs or the need to regulate its temperature (Barghusen, et al., 1997; Lubin and Henschel, 1996; Lubin, et al., 1991; Zevenbergen, et al., 2008).

1.1.5. Other Webs

Despite an abundance and diversity of other spiders, there are relatively few descriptions of the architectural features of their webs. Sheet webs are the three dimensional webs of spiders of the mega-diverse family Linyphiidiae. These webs have an aerial or substrate-bound retreat resembling a dome or sheet with few upward or downward running substrate-attaching lines (Benjamin, et al., 2002; Benjamin and Zschokke, 2002, 2003, 2004; Blackledge, et al., 2009; Eberhard, et al., 2008). There are, unfortunately, few studies of sheet web plasticity (Blackledge, et al., 2011). The few that have been made have generally examined the behavioural responses of Linyphiids feeding on insect pests in agricultural ecosystems, They suggest that, like orb webs, variations in the type and size of prey consumed and spatial supporting structures influence web building behviour and, consequently, web architectures (Benjamin, et al., 2002; Harwood and Obrycki, 2005; Rodríguez, 2000), although there is a lack of specific descriptions of these architectural changes.

1.2. Models Explaining Web Architectural and Performance Plasticity

Researchers have applied various ecological, behavioural or physiological models to their field and laboratory observations. The most commonly applied models are: optimal foraging models, optimal performance models, physiological limitation models, responses to cues, limits to learning models and extended phenotype models.

1.2.1. Optimal Foraging Models

These models posit that the principal fitness benefit of plasticity in web architectural components and whole webs is that it bestows spiders with the ability to cater the performance of their web to capture the most profitable or abundant prey available, assuming prey profitability or abundance can be readily assessed in any environment by the spider at minimal costs to the spider (Blamires, 2010; Blamires, et al., 2011; Higgins and Buskirk, 1992; Miyashita, 1997; Sandoval, 1994; Venner and Casas, 2005; Venner, et al., 2000). The model assumes all of the energy that is spent on foraging is expended building the web. Accordingly, web components are measured as a means to calculate total investment in the web in a currency as a proxy for the total energetic cost, e.g. mm of silk invested. The energetic return (or number or biomass of prey) is calculated as a product of the total prey captured and consumed over a finite time. More recently, nutrients, e.g. proportion of protein, has been used to quantify the value of a particular type of prey (Blamires, et al., 2009; Mayntz, et al., 2003, 2009). While the model can explain why spiders might vary their web architecture when exposed to different prey, it has shortcomings. Firstly, the assumption that all of a spider's foraging expenditure is in web building is probably untrue. Spiders may expend considerable additional energy attacking, capturing and consuming prey, and producing silks for purposes such as decorations and web frames, that may not play any role in foraging. Additionally, many spiders may make a substantial energetic foraging investments in their bodies, for example they may produce pigments that attract prey to their webs (Oxford and Gillespie, 1998; Théry and Casas, 2009). Moreover, each of the components of a web is not of an equal energetic cost to produce. Aggregate and pyriform glues, for instance, are made of a mixture of organic and inorganic low and high molecular weight compounds, most of which are metabolically expensive to synthesize (Higgins and Rankin, 1999; Sahni, et al., 2011; Sahni, et al., 2012; Townley, et al., 2006; Vollrath and Knight, 2001). Radial threads on the other hand are composed of silk that comprises chains of readily synthesizable amino acids such as glycine and alanine (Craig, 2003; Vollrath and Knight, 2001). The number of strands also varies between threads, and strand thickness scales with spider weight, so length of thread may not be a good measure of silk volume. Furthermore, the model seems to ignore the energetic cost of spinning the web (i.e. climbing up and down laying silk into a web). Yet, this cost may be exceptional and vary from spider to spider (Venner, et al., 2003).

1.2.2. Optimal Performance Models

Optimal performance models posit that particular web architectures have the ability to withstand the impacts of particular prey or to withstand particular environmental stresses, e.g. high velocity winds, and suggest that plasticity in specific web components maximizes the web's performance for particular tasks across different environments (Blamires, et al., 2011; Craig, 1987; Cranford, et al., 2012; Liao, et al., 2009; Lin, et al., 1995; Vollrath, et al., 1997). There are performance-based explanations for many co-variation among web components, including the finding that the number of radii invested co-varies with web size or spiral length in orb webs because the vibratory signals that traverse radii are damped by excessive spirals so more radii are deposited to maximize signal clarity (Blamires, et al., 2011; Klarner and Barth, 1982; Landolfa and Barth, 1996; Tarakanova and Buehler, 2012). The ability of webs to withstand the impacts of fast moving and/or heavy (i.e. high kinetic energy) prey across environments builds on the optimal foraging model because it suggests that webs are constantly being redesigned to efficiently catch the most profitable prey available to a spider in a given environment (Blackledge and Eliason, 2007; Blamires, et al., 2011; Cranford, et al., 2012; Tso, et al., 2007; Venner and Casas, 2005). A shortcoming of these models has been when an architectural design appears to have a purpose but experiments reveal that the webs do not perform the supposed purpose. For example, the elongation or condensation of web capture surface areas across habitats has been construed to be a plastic response to optimally capture certain types of prey (Eberhard, 1986a; Nentwig, 1987; Nentwig and Heimer, 1987). Nevertheless, in many instances experiments show that web capture area elongation or condensation is a means to better utilize limited space (Harmer, 2009; Harmer and Herberstein, 2009; Nentwig and Heimer, 1987; Robinson and Lubin, 1979).

1.2.3. Physiological Limitations Models

These models suggest that physiological limitations, e.g. limitations imposed by metabolism (Prestwich, 1977), mean an optimal web type may not be feasible in all conditions so spiders "make the best of a bad lot" and build the most optimal web in a given environment (Blamires, 2010; Higgins, 1992, 2006; Mayntz, et al., 2009; Sherman, 1994). This model may explain web architectural variations across ontogeny (Eberhard, 1986b; Hesselberg, 2010; Higgins, 1992, 2006) and web architecture variations in spiders exposed to thermally extreme or dehydrating environments (Liao, et al., 2009; Lubin and Henschel, 1996; Pulz, 1987). It may also explain why sometimes web architectural components vary at significant costs to web performance (Blackledge and Eliason, 2007; Blamires, et al., 2011). For instance, energetic constraints render it not possible to increase or decrease the size or water content of the aggregate glue to counter the dehydrating effects of drying environments. Orb web spiders therefore compromise by producing fewer spirals packed with more glue, or more spirals with sparsely allocated glue, or by variation in droplet water content (Crews and Opell, 2006; Liao, et al., 2009; Opell, 1997; Opell and Hendricks, 2007; Opell, et al., 2011).

1.2.4. Responses to Cues

This model assumes that spiders have no prior "motivation" for adjusting their web architecture, rather they vary it in response to specific environmental cues or stimuli, e.g. gravity, time of day, volume of glandular silk available, proximity of the body or a specific leg to the auxiliary spirals or radii at a specific time (Eberhard, 1987; Eberhard, 1988; Eberhard and Hesselberg, 2012; Sandoval, 1994; Tso, 2004; Vollrath, 1986). The model is based on findings that (i) experimentally altering specific cues causes alterations to specific architectural components, e.g. placing a spider upside down to reverse the influence of gravity induces reversed web asymmetry in *Cyclosa* spp. (Nakata and Zschokke, 2010; Zschokke and Nakata, 2010), (ii) manipulating silk reserves leads to variations in specific architectural components, e.g. decorations (Tso, 2004), (iii) computer simulations and virtual spiders building webs with specific rules and specific responses inputted design webs with architectures remarkably similar to those built by real spiders (Eberhard, 1969; Krink and Vollrath, 1997, 1999), and (iv) spiders will alter the architecture of their web if fed neurotoxins or when an artificial tactile stimulus is applied to or near the web eliciting specific response behaviours at specific moments during web building (Blamires, et al., 2011; Hesselberg and Vollrath, 2004; Nakata, 2007, 2008; Walcott, 1969). These models ascribe specific behavioural changes to specific stimuli, resulting in web architectural changes (Eberhard, 1987; Eberhard and Hesselberg, 2012; Nakata, 2008). Nonetheless, spiders that respond predictably to stimuli on first encounter may become conditioned to stimuli or will respond differently to specific stimuli in different contexts (Eberhard and Hesselberg, 2012; Heiling and Herberstein, 1999; Nakata, 2007). Thus, the model has limitations in explaining responses to repeated stimuli, variability in web architectural plasticity among individuals under identical conditions, the phenomenon of prior experiences shaping web architectures (Heiling and Herberstein, 1999; Sandoval, 1994; Venner, et al., 2000), and the plastic responses to non-stimulatory cues, such as nutrients (Blamires, 2010; Blamires, et al., 2011; Blamires, et al., 2009; Mayntz, et al., 2003, 2009).

1.2.5. Limits to Learning

These models posit that repeatedly building the same web or decorating the web repeatedly in a similar fashion induces learning in predators and/or prey, which diminishes the functional capacity of the web or web decorations (Blackledge, 1998; Blackledge and Wenzel, 1999; Bruce, et al., 2001; Cheng, et al., 2010; Craig, 1994a; Craig, 1994b; Craig, et al., 2001; Uhl, 2008). According to these models, variation in certain web components should be unpredictable because of inherent variability independent of the environment (Blamires, 2010; Craig, et al., 2001). The model has been used to explain the wide variation in patterns of web decorations found within and between juvenile *Argiope* spp., particularly those that add cruciform decorations to their webs (Bruce, et al., 2001; Cheng, et al., 2010; Craig, 1994a; Craig, 1994b; Craig, et al., 2001; Rao, 2010). Indeed, a laboratory manipulation of *Argiope* web spectral and architectural properties found that stingless bees learn to avoid webs if repeatedly exposed to the same colour and/or pattern (Craig, 1994a; Craig, 1994b). One study (Rao, 2010), however, has revealed contradictory findings. Nonetheless, different types of bees learn colours and shapes at different rates (Dyer, 2012) or the same bees may respond differently to stimuli depending on the time or context of the encounter (Rao, et al., 2008). There are thus many ways that predators and prey learn to recognize spider webs and their signals. Notwithstanding the variations in *Argiope* web decorations induced by variations in prey types, nutrients and web damage (Blamires, 2010; Blamires, et al., 2009; Walter and Elgar, 2011), it is plausible that learning by predators and/or prey drives the inherently high variability in decoration designs in *Argiope* spp. to some extent (Blamires, 2010; Craig, et al., 2001; Uhl, 2008).

1.2.6. Plastic Extended Phenotype Model

This model suggests that a spider web is a complex extension of the spider's phenotype, depicting its foraging effort, ontogeny, size, physiological limitations, sensory strategies, and past interactions with predators and prey (Baba, et al., 2012; Bailey, 2012; Blamires, 2010; Dawkins, 1982; Hansell, 2005; Scharf, et al., 2011). The web, accordingly, dynamically shifts in architecture to optimize the design across a suite of ecological circumstances (Baba, et al., 2012; Blamires, 2010; Higgins and Buskirk, 1992; Nakata, 2012). Silk properties play a role in shaping the optimal design, but the nature and extent of the feedback mechanisms between silk properties, web properties and spider fitness are not well understood (Blackledge, 2012; Blamires, 2010; Harmer, et al., 2011). More complete examinations of the role of the various investments, trade-offs and feedback cycles between the environment, spider phenotypes, extended phenotypes and fitness are required to fully understand the processes by which webs depict the phenotypes of individual spiders (Blamires, 2013).

2. Plasticity in Silk Mechanics

The properties of biomaterials often depend on physical conditions, such as temperature, pH or humidity (Aksakal and Alekberov, 2009; Puthoff, et al., 2010; Taylor, et al., 2004). Furthermore, they may vary among and within individuals of the same species in relation to their internal conditions and/or environmental factors (Caro, et al., 2008; Carrington, 2002; Salem, et al., 1992). Animals may alter the properties of their secreted materials by actively changing the physico-chemical environment of the material (Vincent, 2002). Spider silk, in particular, is known as a material with highly variable properties (Craig, 2003). This makes it a difficult material for polymer scientists to study. Nonetheless, its variability is of intense interest to evolutionary biologists (Craig, 2003; Harmer, et al., 2011). In this review, we will focus on three types of material properties: (i) MAS tensile properties, (ii) MAS supercontraction and (iii) capture spiral silk stickiness.

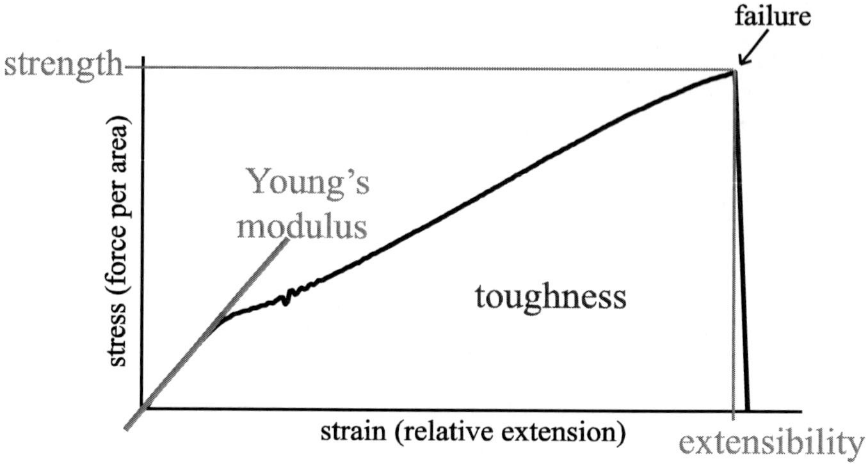

Figure 3. Stress-strain curve of spider silk, with main tensile properties indicated.

2.1. Plasticity in MAS Tensile Behaviour

Tensile properties describe the behaviour of a material when a fiber made of this material is stretched axially. Tensile properties are calculated from stress-strain curves that show the stress (force applied on the fiber per area) as a function of strain (relative elongation of the fiber) (Figure 3). The main tensile properties include:

- strength: the stress the material can sustain before breaking
- extensibility: the strain the material can sustain before breaking
- toughness: the energy the material can absorb before breaking. It is the area under the stress-strain curve
- Young's modulus: a measure of stiffness. It is the slope at the beginning of the stress-strain curve

Specific tensile properties are associated with certain secondary structures. More precisely, strength and stiffness increase with the proportion of β-sheets in the silk. In contrast, extensibility correlates with the presence of helicoidal structures, such as 3_{10} helices or β-turns (Blackledge, et al., 2012; Hayashi, et al., 1999; Hinman, et al., 2000). Toughness depends on strength, extensibility and stiffness, such that it is not associated to a specific molecular structure.

2.1.1. Environmental Parameters Affecting MAS Tensile Behaviour

The silk whose tensile behaviour is best characterized, is MAS, an "all-purpose" silk that is laid by walking spiders and is used in webs. MAS forms the radii and frame of orb webs and is the main silk used in cobwebs (Benjamin and Zschokke, 2002; Blackledge, et al., 2005a; Boutry and Blackledge, 2009; Foelix, 1987; Work, 1976). While MAS is sometimes collected from webs or free-running spiders, it is often obtained by forcibly pulling silk off a restrained animal, a method called "forcible silking". Forcible silking has been used extensively because it limits variability in silk mechanics (Guinea, et al., 2005b). Indeed, MAS is highly variable, particularly its post-yield properties, such as strength and extensibility (Garrido, et al., 2002a). MAS tensile properties vary among individuals, and across time within individuals (Madsen, et al., 1999).

Several factors affecting silk tensile behaviour have been identified. Firstly, the way silk is spun influences silk mechanics. Forcibly obtained MAS is stiffer, stronger and less extensible than silk spun while running (Madsen and Vollrath, 1999). Some of these differences may be due to the the use of CO_2 as an anesthetic rather than the act of forcible silking itself (Work, 1976). However, even when the spider is not under anesthesia, forcibly obtained silk differs from naturally spun silk (Perez-Rigueiro, et al.,

2005). The effect of spinning methods on silk tensile behaviour is most pronounced in orb-weaving species, and weakest in more basal clades such as haplogyne spiders. Hence, for orb-weavers, forcibly-obtained silk is stronger than naturally-spun silk but there is no difference for haplogyne spiders (Boutry, et al., 2011).

Secondly, spiders seem to adjust silk tensile behaviour according to how the silk is used. For instance, silk spun by the banded garden spider *Argiope trifasciata* is stronger when the spider climbs up than when the spider walks on a horizontal surface (Garrido, et al., 2002b). As silk threads are used as safety lines by spiders (Osaki, 1996, 1999), the stronger silk thread spun by climbing spiders may decrease the risk of the spider hurting itself if it falls. Similarly, silk strength, as well as proportion of β-sheets, increases with the height at which silk is spun, and is even higher if the spider drops instead of climbing (Pan, et al., 2004). Another example involves silk tensile properties in different regions of the common house spider *Achaearenea tepidariorum*'s cobweb. Silk used in the supporting threads is generally strong while silk used in the gumfooted threads is more compliant (Boutry and Blackledge, 2009). These differences may improve cobweb function by allowing it to support heavy prey and better distributing loads among threads.

Finally, prey types and foraging success may influence silk mechanics. Different prey types vary in nutritional value and provide different stimuli to the spider, which may explain the variation in silk tensile behaviour with diet (Blamires, et al., 2010a; Blamires, et al., 2012a). However, different experiments, using a variety of procedures, have produced contradictory results. As such the precise means by which prey influence silk properties remains unclear. For instance, the giant wood spiders *Nephila pilipes*, fed large prey (crickets), spun thicker but more compliant silk threads than spiders fed small prey (flies) (Tso, et al., 2007). However, in another study, common house spiders, *Achaearanea tepidariorum*, receiving fast, heavy prey made stronger silk than spiders receiving smaller, slower prey (Boutry and Blackledge, 2008).

The results from this last experiment could be due to differences in spider weight or condition rather than prey type. Indeed, spiders loaded with added weights will spin thicker threads made of more compliant silk (Vollrath and Köhler, 1996).

Furthermore, prey nutritional value and different stimuli on the web may interact to affect silk tensile properties (Blamires, et al., 2010a). It is clear that the spider's foraging success has an influence on silk tensile mechanics. The golden orb weaver, *Nephila clavipes*, produces stiffer and less extensible silk when starved than when well fed (Guehrs, et al., 2008). Another experiment in the same species confirmed that deprivation of the amino acids alanine, glycine, tyrosine and arginine leads to the production of silk with lower ultimate strength and extensibility (Zax, et al., 2004). For spiders deprived of alanine and glycine, the decrease in tensile performance is accompanied by fewer β-sheets in the silk.

Plasticity in silk tensile properties is well documented, particularly for MAS, yet the mechanisms of plasticity are unclear. Two main hypotheses can explain plasticity in MAS tensile behaviour: (i) changes in silk chemical composition and (ii) spinning processes. There is evidence supporting both hypotheses, suggesting that they are both involved in plasticity.

2.1.2. Changes in MAS Chemical Composition

The first hypothesis explaining plasticity in MAS tensile behaviour states that variations in amino acid composition drives variations in tensile properties. Different amino acids form different secondary structures, and thus confer various tensile properties to the silk (Figure 4). Series of alanine amino acids, alone (A_n, where A is alanine) or in association with glycine (GA_n, where G is glycine), form β-sheets, thus making silk strong and stiff (Jelinski, et al., 1999; Jenkins, et al., 2010; Kümmerlen, et al., 1996; Simmons, et al., 1994; Xu and Lewis, 1990). Proline is found in GPGXX motifs, where P is proline and X is one of a small group of amino acids. These proline-rich motifs likely form β-turns, making silk extensible and compliant (Hayashi and Lewis, 1998; Hayashi, et al., 1999; Ohgo, et al., 2006). Finally, glycine is also found in GGX motifs, which form 3_{10} helices (Bram, et al., 1997; Kümmerlen, et al., 1996; van Beek, et al., 2002). These helices should confer a certain compliance to the silk (Dong, et al., 1991; Scheibel, 2004), as well as extra strength (Blackledge, et al., 2012). In general, proteins containing proline and glycine above a certain threshold are elastomeric, while those below another threshold are amyloidogenic (Rauscher, et al., 2006). This appears to be the case in silk too, as the proportion of glycine in spider silk correlates with the level of disorder (Dicko, et al., 2007).

Variation in amino acid composition may be due to changes in the ratio of silk proteins. Silk of orbicularian spiders is made of two proteins, MaSp1 and MaSp2, which differ in their amino acid composition. MaSp1 is rich in alanine and glycine, while MaSp2 contains the amino acid proline, which is not found in MaSp1 (Rising, et al., 2005). Thus, variations in the ratio of MaSp1 to MaSp2 may alter MAS amino acid composition, and by extension, MAS tensile behaviour. Bioenergetic analysis has shown that MaSp2 is more costly to synthesize than MaSp1 (Craig, et al., 1999). This may be why starving spiders or spiders deprived of protein decrease their expression of MaSp2, yielding stiffer silk (Blamires and Tso, 2013; Blamires, et al., 2012c; Guehrs, et al., 2008). Furthermore, several alleles have been found for the MaSp1 gene (Ayoub and Hayashi, 2008; Rising, et al., 2007; Tai, et al., 2004). This means that variation in amino acid composition could be due to the expression of one allele or the other in heterozygous individuals. Finally, one may try to explain changes in amino acid composition by changes in the protein sequence, for instance resulting from alternative splicing. However, MaSp genes are made of a single, large exon (Ayoub, et al., 2007). Therefore, alternative splicing cannot take place. However, to date, our knowledge of MaSp sequences remains very limited. We only know the full sequence from both MaSp of one species, and we are still unsure whether MaSp1 and MaSp2 are both present in many clades. Therefore, the three mechanisms of amino acid variation presented here remain highly hypothetical.

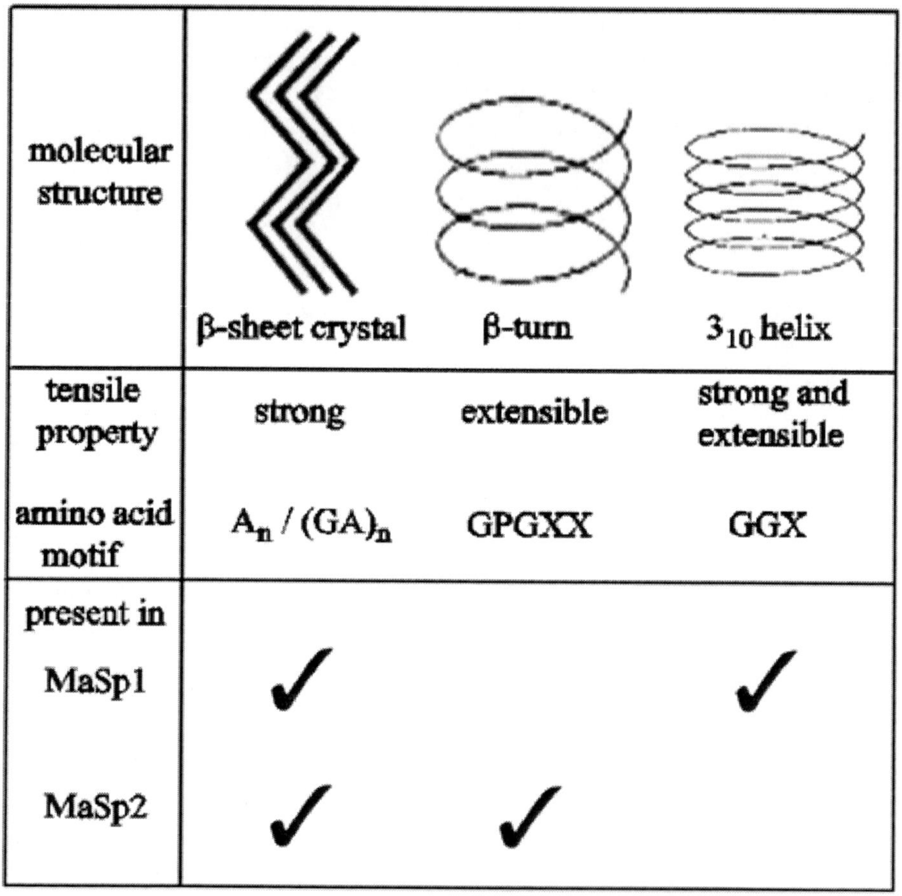

Figure 4. Hypothesized relationship between amino acid motifs, silk secondary structures and tensile properties in the two MAS proteins, MaSp1 and MaSp2. A represents alanine, G, glycine, P, proline and X, one of a small subset of amino acids.

Regardless of its mechanism, MAS amino acid composition differ between individuals of the same species (Blamires, et al., 2010a; Craig, et al., 2000; Tso, et al., 2005), which may explain why there are intraspecific differences in MAS tensile behaviour. In the lab, a comparison of silk spun by *N. pilipes* fed crickets versus *N. pilipes* fed flies found differences in amino acid composition among individuals (Tso, et al., 2007; Tso, et al., 2005). These differences correlated with differences in silk secondary structures (Tso, et al., 2005) and tensile properties (Tso, et al., 2007). Protein deprivation in three orb-weaving species resulted in spiders making silk with lower content of proline and glutamine, two amino acids that are costly to synthesize (Blamires, et al., 2012c). However, in another study, *Nephila clavipes* deprived of specific amino acid all spun silk with the same chemical composition, but with different tensile behaviour, suggesting multiple mechanisms may induce MAS plasticity (Zax, et al., 2004).

Other studies suggest that plasticity in MAS tensile behaviour is independent of amino acid variation. Tensile properties plasticity without changes in silk chemistry was observed in *Cyclosa mulmeinensis* (Liao, et al., 2009). Furthermore, a wide range of MAS amino acid compositions are found across species but without an equally wide range of MAS tensile properties (Blamires, et al., 2012a). All these results point to the existence of another mechanism explaining plasticity in MAS tensile behaviour.

2.1.3. Spinning Effects

The second hypothesis explaining plasticity in the tensile behaviour of MAS suggests that the processing conditions during spinning influence tensile properties. The secreted silk is stored within silk glands as a liquid dope. As this solution goes through the spinning duct, it experiences changes in pH, ion and water content and is subjected to friction before its extrusion as a solid fiber (Askarieh, et al., 2010; Dicko, et al., 2004a; Dicko, et al., 2004b; Knight, et al., 2000; Knight and Vollrath, 2001). These biochemical and mechanical processes are collectively known as "spinning processes", and they may affect the molecular structure and tensile behaviour of the silk at any stage of spinning (Kojić, et al., 2006).

The higher strength and stiffness of forcibly-obtained MAS may be due to spiders applying 40 times as much friction during forcible silking as they do when freely laying silk (Ortlepp and Gosline, 2004). When laying silk freely, spiders apply different tensions on the silk thread, explaining the high variability of silk tensile properties. In contrast, the tension applied is relatively constant after just a few minutes during forcible silking (Ortlepp and Gosline, 2004). This could explain why forcibly-obtained MAS displays less variation than naturally-spun silk (Guinea, et al., 2005b). Spiders possessing a well-developed spinning duct valve, e.g. orb-weavers, may be better at controlling MAS reeling speed and resisting forcible silking than spiders lacking a complex valve, e.g. haplogyne spiders (Guinea, et al., 2005b). This ability to "brake" during spinning can explain differences in plasticity between spider clades (Boutry, et al., 2011). Experiments have confirmed that tension applied during silk spinning correlates positively with silk stiffness and strength (Perez-Rigueiro, et al., 2005).

A second parameter affecting silk tensile behaviour is how quickly silk is drawn. Using thermal analysis, Guess and Viney (Guess and Viney, 1998) showed that faster reeling speeds lead to more compliant silk. Nonetheless, other measurements have contradicted this conclusion. Vollrath *et al.* later found that increasing reeling speeds by ~100 mm/s produces stiffer, stronger and less extensible MAS (Vollrath, et al., 2001). The secondary structures of MAS also vary with spinning speed, with the proportion of β-sheets formed peaking at 20 mm/s reeling speed (Pan and Liu, 2009). Faster reeling-speed probably increases the molecular orientation in the silk fiber, explaining its increased strength and stiffness (Knight and Vollrath, 1999).

Another theory regarding MAS tensile plasticity involves the concept of a supercontracted "ground" state. Supercontraction is the tendency of MAS to shrink up to 50% of its original length and become rubbery when exposed to water (Work, 1977). According to the theory, supercontracted silk is as compliant, extensible and elastomeric as possible (Elices, et al., 2011). As the silk within the duct is wet, it should be supercontracted. The amount of strain the wet silk sustains inside the duct would thus determine the tensile properties of the end product (Guinea, et al., 2005a). This theory is supported by experiments involving silk spun in different media. Silk spun into water under tension is stiffer and stronger than silk spun in air under the same tension (Liu, et al., 2005). While the study concluded that water itself improved silk tensile properties, it also supports the idea that it is the tension applied on the wet silk that improved its performance. Indeed, stretching silk in water after spinning improves tensile performance by promoting the formation of β-sheets (An, et al., 2011).

2.2. MAS Response to Water: Supercontraction

An interesting case of MAS plasticity is in its response to varying humidity. When exposed to humidity levels above ~70%, MAS fibers supercontract, shrinking by up to half their length (Boutry and Blackledge, 2010; Work, 1977). In restrained fibers, as the thread is unable to shrink, humidity instead leads to large tensions. These tensions can reach up to 150 MPa or 10% of the average MAS strength (Boutry and Blackledge, 2010). Silk tensile properties are extremely different between native (i.e. non-supercontracted) and supercontracted silk (Blamires, et al., 2012a). Unrestrained supercontracted silk is up to 1000 times more compliant and twice as extensible as native silk (Savage and Gosline, 2008a).

In general, silk becomes more compliant and extensible when humidity increases, even at humidity levels below supercontraction threshold (Vehoff, et al., 2007). Unrestrained supercontracted silk behaves like a rubber (Gosline, et al., 1984). This leads to lower elastic hysteresis, which means that supercontracted silk stores energy much more than native silk (Savage and Gosline, 2008b; C. Boutry, V. Sahni and T. A. Blackledge, unpublished). Recently, it was found that the level of supercontraction depends on the rate at which humidity changes. A change from 30% to 90% humidity in five minutes results in half as much tension in restrained MAS fibers as the same change taking place in one minute (Agnarsson, et al., 2009). Extensive research has been conducted on the mechanisms of supercontraction. The following overview is not exhaustive, but covers the main findings on this topic.

To understand supercontraction, we first need to detail silk molecular structure. Silk is often modeled as β-sheet crystals embedded in a matrix (Grubb and Jelinski, 1997; Warwicker, 1960). The matrix contains various structures, likely including 3_{10} helices, β-turns or spirals and non-crystalline β-sheet (Bram, et al., 1997; Hayashi and Lewis, 1998; Jenkins, et al., 2010; Kümmerlen, et al., 1996; Ohgo, et al., 2006). When the silk is dry, the structures of the matrix are held together by hydrogen bonds, forming a hard network (Figure 5).

Supercontraction affects the matrix by disrupting hydrogen bonds (Eles and Michal, 2004; Grubb and Ji, 1999; Termonia, 1994; Work and Morosoff, 1982). As a result, the molecular structures of the matrix are free to rearrange (Yang, et al., 2000).

Following thermodynamic laws, the structures lose their orientation, resulting in a shrinking of the fiber (Grubb and Ji, 1999; Parkhe, et al., 1997). Because the structures of the matrix are no longer held together, the silk can extend easily, explaining the low stiffness and high extensibility of supercontracted silk (Ene, et al., 2010). The association between amount of proline in the silk and level of supercontraction confirms that β-turns formed by GPGXX motifs are involved in supercontraction (Blamires, et al., 2012a; Creager, et al., 2010; Liu, et al., 2008; Savage and Gosline, 2008a).

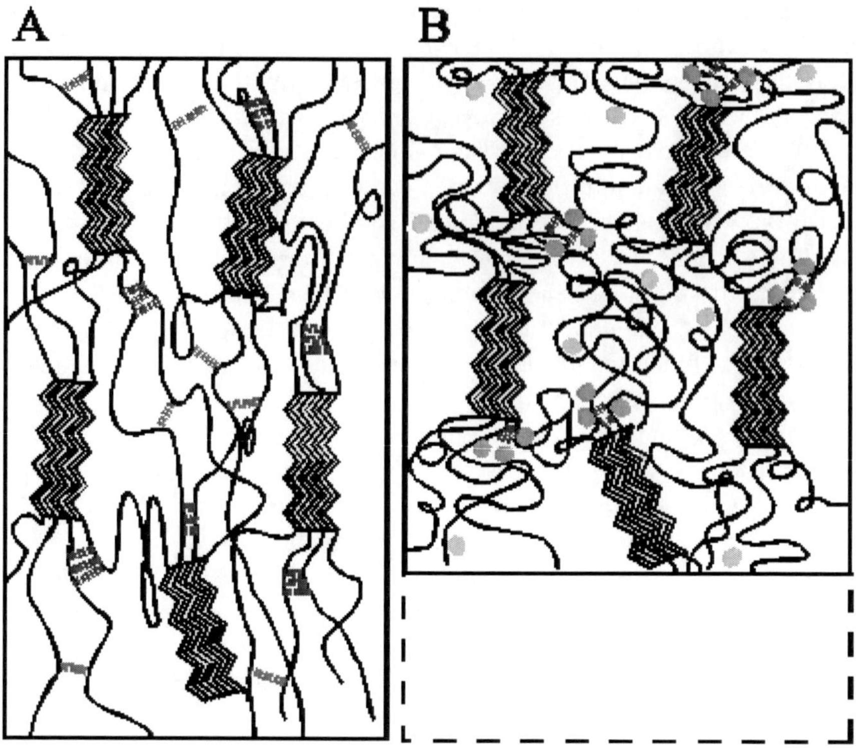

Figure 5. Molecular structure of native (A) and supercontracted (B) spider silk. Native silk is made of β-sheet crystals (wavy rectangles) embedded in a matrix rich in helicoidal structures (solid lines) held parallel by hydrogen bonds (dotted lines). Water molecules (circles) disrupt the hydrogen bonds, resulting in a rearrangement of the helicoidal structures and a shrinking of the fiber.

2.3. Plasticity in Capture Spiral Stickiness

The stickiness of the capture spirals of orb webs is also highly plastic. Stickiness is usually measured as the force per area needed to detach a surface stuck to the silk. Silk stickiness arises from wet glue in ecribellate spiders or Velcro® -like "wooly" fibrils in cribellate spiders (Opell, 1997) (Figure 6). Just like the tensile behaviour of MAS, the stickiness of the capture spiral is affected by starvation and/or nutrients, silk use, ontogeny and environmental conditions. The density of glue decreases during starvation, resulting in stickiness dropping by half in less than a month (Crews and Opell, 2006).

Stickiness also varies within the orb-web: outer areas of the web are stickier than inner areas, maybe because silk reserves get depleted during web construction (Opell, et al., 2009). Furthermore, silk stickiness varies through the season, possibly because resources are re-allocated from web construction to reproduction in the fall (Opell, et al., 2009).

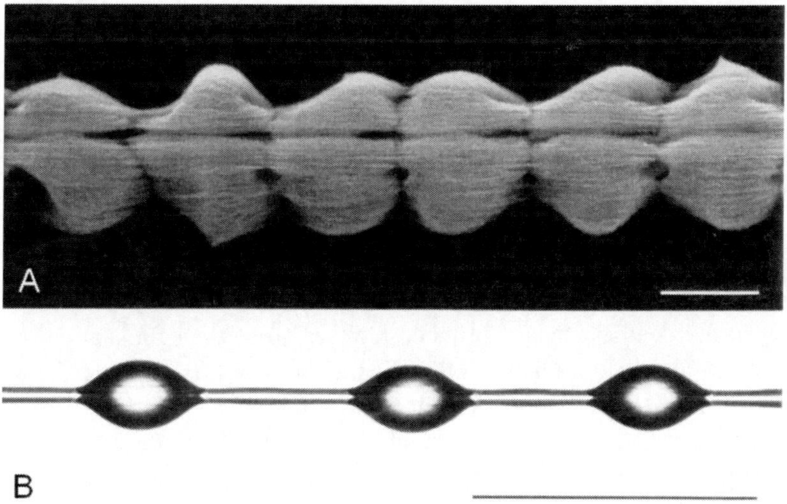

Figure 6. Cribellate (A) and ecribellate (B) silk threads. Reproduced from Opell and Bond, 2001. Scale bars are 100 μm.

Environmental humidity and wind speed affects web stickiness, at least for ecribellate silk threads. Water deposition on capture threads increases the volume of glue droplets (Edmonds and Vollrath, 1992). However, droplet volume may not correlate with stickiness (Wu, Blamires, Wu & Tso, submitted). Water also increases silk thread and droplet extension, increasing the number of droplets that stick to an object and likely increasing adhesion (Opell, et al., 2011). However, humidity does not affect stickiness in cobweb-weavers, such as *L. hesperus*, suggesting different glue compositions between orb- and cobweb-weaving spiders (Sahni, et al., 2011). Water also improves stickiness for cribellate silk threads spun by entelegyne spiders (Hawthorn and Opell, 2002). This suggests hygroscopic forces are involved in cribellate silk adhesion (Opell, 1995). Entelegyne spiders threads differ from haplogyne spiders threads by including nodes that may be hydrophilic.

In addition to glue morphology, the composition of the glue may also vary and affect stickiness. Capture spiral glue is made of low molecular weight components (Townley, et al., 1991). The type and quantity of these components change in function of spiders diet. Upon starvation, the proportion of less readily synthesized compounds, particularly choline and glycine betaine, decreases (Townley, et al., 2006). This suggests that a change in thread stickiness may be associated with starvation. A decrease in stickiness is also observed when spiders are deprived of proteins without any deprivation of caloric intake (S.J. Blamires, V. Sahni, A. Dhinojwala, T.A. Blackledge & I.M. Tso, unpublished). Spiral glue chemical compositions vary among individuals of the same species (Higgins, et al., 2001). This variation may be induced by diet, but also by other factors, such as genetic and ontogenetic differences between spiders as well as adjustment to environmental conditions (Higgins, et al., 2001). Spiders may be able to adjust silk stickiness to their environment. For instance, the giant orb weaver *Nephila pilipes*, has stickier webs when fed dead crickets than when fed live crickets or flies, alive or dead (Blamires, et al., 2011). On the other hand, there is no evidence of genetic differences in silk glue composition or stickiness between spiders. Two genes coding for capture spiral glue have been identified in *Nephila clavipes* but we do not know whether there are different alleles for these genes (Choresh, et al., 2009). Furthermore, the stickiness of spiral and gumfooted threads depends on the morphological features of the prey adhering to the web. In general, both ecribellate and cribellate spider threads adhere better to insects with smooth surfaces (Eisner, et al., 1964; Opell, 1994; Opell and Schwend, 2007).

3. Combined Effects of Web Architecture and Silk Properties on Web Performance

The overall performance of a web across environments is correlated with the expression of the individual architectural components and silk properties across those same environments (Blamires, 2010; Blamires, et al., 2011; Boutry and Blackledge, 2009; Denny, 1976; Harmer, et al., 2011; Higgins and Buskirk, 1992; Miyashita, 1997; Sandoval, 1994; Tso, et al., 2007). However, our understanding of how the different web components combine to influence overall web performance is still limited (Harmer, et al., 2011). Plasticity in web architecture may correlate or be compensated by plasticity in silk mechanics in ways that improve web performance. Aspects of web performance include, among other factors, the ability to absorb impact, to retain prey through thread adhesion and to detect prey though transmission of prey vibrations (Blackledge and Zevenbergen, 2006; Masters, et al., 1986).

Of the three, the ability of webs to absorb impact has received the most attention. The impact absorbing potential of the web co-varies with the number, or width, of radial threads (Blamires, et al., 2011; Cranford, et al., 2012; Ko and Jovicic, 2004; Sensenig, et al., 2012; Tso, et al., 2007; Venner, et al., 2000). Concomitantly, impact absorbing potential depends on silk nonlinear material behaviour and silk energy absorbing capacity in the plastic range (Cranford, et al., 2012; Ko and Jovicic, 2004). Furthermore, the threads within the web are laid under tension (Wirth and Barth, 1992). This tension makes the web more resistant to damage and increases the frequency of web oscillations after prey impact (Alam, et al., 2007). Therefore, variation in either architecture or silk mechanics will affect web's ability to sustain impacts. For instance, the softening and increased elasticity of wet silk leads to capture success doubling in *Argiope* orb webs exposed to high humidity (C. Boutry and T. A. Blackledge, unpublished).

Likewise, web's ability to retain prey depends on both capture spiral stickiness and web architecture. The prey retention capacity of the capture spirals co-varies with capture thread length and spacing, chemical properties of the glue droplets and extensibility of the flagelliform axial fiber (Opell, 1999; Opell, et al., 2011; Opell, et al., 2009; Sahni, et al., 2011; Tarakanova and Buehler, 2012), but also with number of radii (Blamires, et al., 2011; Denny, 1976; Tarakanova and Buehler, 2012). The stickiness of cobweb and orb web glues depends on the physical and chemical nature of the constituent flagelliform (MAS in the case of cobwebs) and aggregate silks (Opell, 1999; Opell and Bond, 2001; Opell and Hendricks, 2007; Opell, et al., 2011; Sahni, et al., 2011). Spiders may adjust stickiness to environmental conditions by changing either web architecture or silk properties. For example, the spiral threads of the dust spider, *Cyclosa mulmienensis*, dehydrate and lose stickiness rapidly when exposed to wind. This spider does not compensate for this performance constraint by changing the properties of its sticky silk, but by changes in its web architecture and/or by aggregating its webs (Wu, Blamires, Wu & Tso, submitted). Stickiness increases with environmental humidity (see "Plasticity in capture spiral stickiness" section). Accordingly webs in relatively low humidity environments will be less sticky than architecturally identical webs in relatively high humidity environments. Strong wind probably has a similar effect of reducing web stickiness (Lin, et al., 1995).

Webs' ability to transmit prey vibration has received little attention to date, even though the vibrations emitted by a struggling prey are the main method of prey detection by spiders (Masters, et al., 1986). Vibration transmission depends on silk mechanics: it increases with silk stiffness and tension (Frohlich and Buskirk, 1982; Walcott, 1963). This may explain why the stiff radii transmit vibrations better than the softer capture spiral (Landolfa and Barth, 1996). Certain web decorations, such as the spiral-shaped decorations of *Cyclosa* spp., may affect silk tension, and by extension, the spider's ability to detect prey (Watanabe, 2000). Although humidity affects silk mechanics, vibration transmission appears similar in *Nephila* webs exposed to high and low humidity (Landolfa and Barth, 1996). However, *Nephila* silk responds to humidity relatively weakly (Boutry and Blackledge, 2010), such that this conclusion may not necessarily extend to other species. Vibration transmission also depends on web architecture. For instance, vibration transmission along radii decreases every time the radius connects with the capture spiral or the auxiliary spiral, such that vibration transmission in webs decreases with the number of rows of spiral (Landolfa and Barth, 1996; Masters, 1984).

CONCLUSION

Spider webs are complex structures whose performance depends on their architecture, the mechanics of the silk that composes them and the interaction of architecture and silk mechanics. Furthermore, we are only beginning to realize that web architecture and silk mechanics are highly plastic. Spiders appear to readily change web architecture and silk mechanical behavior as a function of their external environment and internal conditions. The influence of plasticity in spider web components on web performance is a topic that is yet to be thoroughly investigated. Yet, it appears to be a key evolutionary driver of spider silk and spider web diversity.

ACKNOWLEDGMENTS

S.J.B. received a National Science Council of Taiwan postdoctoral fellowship (NSC-98-2811-B-029-002) while compiling this review. S.J.B. would like to thank Chen-Pan Liao for assistance with the figures.

REFERENCES

Agnarsson, I., Boutry, C., Wong, S.C., Baji, A., Dhinojwala, A., Sensenig, A.T. and Blackledge, T.A. (2009). Supercontraction forces in spider dragline silk depend on hydration rate. *Zoology*, 112, 325-331.

Agnarsson, I. and Coddington, J.A. (2006). Notes on web and web plasticity and description of the male of *Achaearanea hieroglyphica* (Mello-Leitao) (Araneae, Theridiidae). *Journal of Arachnology*, 34, 638-641.

Aksakal, B. and Alekberov, V. (2009). The effect of temperature and water on the mechanical properties of wool fibres investigated with different experimental methods *Fibers and Polymers*, 10, 673-680.

Alam, M.S., Wahab, M.A. and Jenkins, C.H. (2007). Mechanics in naturally compliant structures. *Mechanics of Materials*, 39, 145-160.

An, B., Hinman, M.B., Holland, G.P., Yarger, J.L. and Lewis, R.V. (2011). Inducing β-Sheets formation in synthetic spider silk fibers by aqueous post-spin stretching. *Biomacromolecules*, 12, 2375-2381.

Andersen, S.O. (1970). Amino acid composition of spider silks. *Comparative Biochemistry and Physiology* 35, 705-711.

Askarieh, G., Hedhammar, M., Nordling, K., Saenz, A., Casals, C., Rising, A., Johansson, J. and Knight, S.D. (2010). Self-assembly of spider silk proteins is controlled by a pH-sensitive relay. *Nature*, 465, 236-U125.

Auld, J.R., Agrawal, A.A. and Relyea, R.A. (2009). Re-evaluating the costs and limits of adaptive phenotypic plasticity. *Proceedings of the Royal Society B: Biological Sciences*,

Ayoub, N.A., Garb, J.E., Tinghitella, R.M., Collin, M.A. and Hayashi, C.Y. (2007). Blueprint for a high-performance biomaterial: Full-length spider dragline silk genes. *PloS one*, 2, 154-166.

Ayoub, N.A. and Hayashi, C.Y. (2008). Multiple recombining loci encode MaSp1, the primary constituent of dragline silk, in widow spiders (Latrodectus: Theridiidae). *Molecular Biology and Evolution*, 25, 277-286.

Baba, Y. and Miyashita, T. (2006). Does individual internal state influence the presence of a barrier web in *Argiope bruennichi* (Araneae: Araneidae)? *Journal of Ethology*, 24, 75-78.

Baba, Y.G., Osada, Y. and Miyashita, T. (2012). The effect of host web complexity on prey-stealing success in a kleptoparasitic spider mediated by locomotor ability. *Animal Behaviour*, 83, 1261-1268.

Bailey, N.W. (2012). Evolutionary models of extended phenotypes. *Trends in Ecology & Evolution*, 27, 561-569.

Barghusen, L.E., Claussen, D.L., Anderson, M.S. and Bailer, A.J. (1997). The effects of temperature on the web-building behaviour of the common house spider, *Achaearanea tepidariorum*. Functional Ecology, 11, 4.

Barrantes, G. and Weng, J.L. (2006). Viscid globules in webs of the spider *Achaeranea tesselata* (Araneae: Theridiidae). *Journal of Arachnology,* 34, 480-482.

Benjamin, S.P., Düggelin, M. and Zschokke, S. (2002). Fine structure of sheet-webs of *Linyphia triangularis* (Clerck) and *Microlinyphia pusilla* (Sundevall), with remarks on the presence of viscid silk. *Acta Zoologica*, 83, 49-59.

Benjamin, S.P. and Zschokke, S. (2002). Untangling the tangle-web: Web construction behavior of the comb-footed spider *Steatoda triangulosa* and comments on phylogenetic implications (Araneae: Theridiidae). *Journal of Insect Behavior*, 15, 791-809.

Benjamin, S.P. and Zschokke, S. (2003). Webs of theridiid spiders: Construction, structure and evolution. *Biological Journal of the Linnean Society,* 78, 293-305.

Benjamin, S.P. and Zschokke, S. (2004). Homology, behaviour and spider webs: Web construction behaviour of *Linyphia hortensis* and *L. triangularis* (Araneae: Linyphiidae) and its evolutionary significance. *Journal of Evolutionary Biology*, 17, 120-130.

Blackledge, T.A. (1998). Signal conflict in spider webs driven by predators and prey. *Proceedings of the Royal Society of London. Series B: Biological Sciences,* 265, 1991-1996.

Blackledge, T.A. (2011). Prey capture in orb weaving spiders: Are we using the best metric? *Journal of Arachnology*, 39, 205-210.

Blackledge, T.A. (2012). Spider silk: A brief review and prospectus on research linking biomechanics and ecology in draglines and orb webs. *Journal of Arachnology*, 40, 1-12.

Blackledge, T.A. and Eliason, C.M. (2007). Functionally independent components of prey capture are architecturally constrained in spider orb webs. *Biology Letters*, 3, 456-458.

Blackledge, T.A. and Gillespie, R.G. (2002). Estimation of capture areas of spider orb webs in relation to asymmetry. *Journal of Arachnology*, 30, 70-77.

Blackledge, T.A. and Hayashi, C.Y. (2006). Silken toolkits: Biomechanics of silk fibers spun by the orb web spider *Argiope argentata* (Fabricius 1775). *Journal of Experimental Biology*, 209, 2452-2461.

Blackledge, T.A., Kuntner, M. and Agnarsson, I. (2011). The form and function of spider orb webs: Evolution from silk to ecosystems. *Advances in Insect Physiology*, 41, 175-262.

Blackledge, T.A., Pérez-Rigueiro, J., Plaza, G.R., Perea, B., Navarro, A., Guinea, G.V. and Elices, M. (2012). Sequential origin in the high performance properties of orb spider dragline silk. *Scientific Reports*, 2, 782.

Blackledge, T.A., Scharff, N., Coddington, J.A., Szuts, T., Wenzel, J.W., Hayashi, C.Y. and Agnarsson, I. (2009). Reconstructing web evolution and spider diversification in the molecular era. *Proceedings of the National Academy of Sciences of the United States of America,* 106, 5229-5234.

Blackledge, T.A., Summers, A.P. and Hayashi, C.Y. (2005a). Gumfooted lines in black widow cobwebs and the mechanical properties of spider capture silk. *Zoology*, 108, 41-46.

Blackledge, T.A., Swindeman, J.E. and Hayashi, C.Y. (2005b). Quasistatic and continuous dynamic characterization of the mechanical properties of silk from the cobweb of the black widow spider *Latrodectus hesperus*. *Journal of Experimental Biology*, 208, 1937-1949.

Blackledge, T.A. and Wenzel, J.W. (1999). Do stabilimenta in orb webs attract prey or defend spiders? *Behavioral Ecology*, 10, 372-376.

Blackledge, T.A. and Wenzel, J.W. (2001a). Silk mediated defense by an orb web spider against predatory mud-dauber wasps. *Behaviour*, 138, 155-171.

Blackledge, T.A. and Wenzel, J.W. (2001b). State-determinate foraging decisions and web architecture in the spider Dictyna volucripes (Araneae Dictynidae). *Ethology Ecology & Evolution*, 13, 105-113.

Blackledge, T.A. and Zevenbergen, J.M. (2006). Mesh width influences prey retention in spider orb webs. *Ethology*, 112, 1194-1201.

Blackledge, T.A. and Zevenbergen, J.M. (2007). Condition-dependent spider web architecture in the western black widow, *Latrodectus hesperus*. *Animal Behaviour*, 73, 855-864.

Blamires, S.J. (2010). Plasticity in extended phenotypes: Orb web architectural responses to variations in prey parameters. *Journal of Experimental Biology,* 213, 3207-3212.

Blamires, S.J. (2013). Spider webs as extended phenotypes. In: Eds, Spiders: morphology, behavior and geographic distribution. New York: Nova.

Blamires, S.J., Chao, I.C. and Tso, I.M. (2010a). Prey type, vibrations and handling interactively influence spider silk expression. *Journal of Experimental Biology*, 213, 3906-3910.

Blamires, S.J., Chao, Y.-C., Liao, C.-P. and Tso, I.M. (2011). Multiple prey cues induce foraging flexibility in a trap-building predator. *Animal Behaviour,* 81, 955-961.

Blamires, S.J., Hochuli, D.F. and Thompson, M.B. (2009). Prey protein influences growth and decoration building in the orb web spider Argiope keyserlingi. *Ecological Entomology*, 34, 545-550.

Blamires, S.J., Lee, Y.H., Chang, C.M., Lin, I.T., Chen, J.A., Lin, T.Y. and Tso, I.M. (2010b). Multiple structures interactively influence prey capture efficiency in spider orb webs. *Animal Behaviour*, 80, 947-953.

Blamires, S.J., Thompson, M.B. and Hochuli, D.F. (2007). Habitat selection and web plasticity by the orb spider *Argiope keyserlingi* (Argiopidae): Do they compromise foraging success for predator avoidance? *Austral Ecology*, 32, 551-563.

Blamires, S.J. and Tso, I.M. (2013). Nutrient mediated architectural plasticity of a predatory trap. *PloS one*, 8, in press.

Blamires, S.J., Wu, C.-L., Blackledge, T.A. and Tso, I.-M. (2012a). Post-secretion processing influences spider silk performance. *Journal of the Royal Society Interface*, 9, 2479-2487.

Blamires, S.J., Wu, C.-L., Blackledge, T.A. and Tso, I.M. (2012b). Environmentally induced post-spin property changes in spider silks: Influences of web type, spidroin composition and ecology. *Biological Journal of the Linnean Society*, 106, 580-588.

Blamires, S.J., Wu, C.-L. and Tso, I.M. (2012c). Variation in protein intake induces variation in spider silk expression. *PloS one,* 7, e31626.

Boutry, C. and Blackledge, T.A. (2008). The common house spider alters the material and mechanical properties of cobweb silk in response to different prey. *Journal of Experimental Zoology A*, 309, 542-552.

Boutry, C. and Blackledge, T.A. (2009). Biomechanical variation of silk links spinning plasticity to spider web function. *Zoology*, 112, 451-460.

Boutry, C. and Blackledge, T.A. (2010). Evolution of supercontraction in spider silk: structure–function relationship from tarantulas to orb-weavers. *Journal of Experimental Biology,* 213, 3505-3514.

Boutry, C., Řezáč, M. and Blackledge, T.A. (2011). Plasticity in major ampullate silk production in relation to spider phylogeny and ecology. *PloS one*, 6, e22467.

Bram, A., Branden, C.I., Craig, C., Snigireva, I. and Riekel, C. (1997). X-ray diffraction from single fibres of spider silk. *Journal of applied crystallography*, 30, 390-392.

Bruce, M.J. (2006). Silk decorations: Controversy and consensus. *Journal of Zoology,* 269, 89-97.

Bruce, M.J., Herberstein, M.E. and Elgar, M.A. (2001). Signalling conflict between prey and predator attraction. *Journal of Evolutionary Biology*, 14, 786-794.

Caro, A.U., Escobar, J., Bozinovic, F., Navarrete, S.A. and Castilla, J.C. (2008). Phenotypic variability in byssus thread production of intertidal mussels induced by predators with different feeding strategies. *Marine Ecology Progress Series*, 372, 127-134.

Carrington, E. (2002). Seasonal variation in the attachment strength of blue mussels: causes and consequences. *Limnology and Oceanography*, 1723-1733.

Cheng, R.-C., Yang, E.-C., Lin, C.-P., Herberstein, M.E. and Tso, I.-M. (2010). Insect form vision as one potential shaping force of spider web decoration design. *The Journal of Experimental Biology,* 213, 759-768.

Choresh, O., Bayarmagnai, B. and Lewis, R.V. (2009). Spider web glue: two proteins expressed from opposite strands of the same DNA sequence. *Biomacromolecules*, 10, 2852-2856.

Coslovsky, M. and Zschokke, S. (2009). Asymmetry in orb-webs: An adaptation to web building costs? *Journal of Insect Behavior,* 22, 29-38.

Craig, C. (1994a). Predator foraging behavior in response to perception and learning by its prey: interactions between orb-spinning spiders and stingless bees. *Behavioral Ecology and Sociobiology,* 35, 45-52.

Craig, C.L. (1987). The ecological and evolutionary interdependence between web architecture and web silk spun by orb web weaving spiders. *Biological Journal of the Linnean Society,* 30, 135-162.

Craig, C.L. (1994b). Limits to learning: Effects of predator pattern and colour on perception and avoidance-learning by prey. *Animal Behaviour,* 47, 1087-1099.

Craig, C.L. (2003). Spiderwebs and Silk: Tracing Evolution from Molecules to Genes to Phenotypes New York: Oxford University Press.

Craig, C.L., Hsu, M., Kaplan, D. and Pierce, N.E. (1999). A comparison of the composition of silk proteins produced by spiders and insects. *International Journal of Biological Macromolecules*, 24, 109-118.

Craig, C.L., Riekel, C., Herberstein, M.E., Weber, R.S., Kaplan, D. and Pierce, N.E. (2000). Evidence for diet effects on the composition of silk proteins produced by spiders. *Molecular Biology and Evolution,* 17, 1904-1913.

Craig, C.L., Wolf, S., Davis, J., Hauber, M. and Maas, J. (2001). Signal polymorphism in the web-decorating spider *Argiope argentata* is correlated with reduced survivorship and the presence of stingless bee, its primary prey. *Evolution*, 55, 986-993.

Cranford, S.W., Tarakanova, A., Pugno, N.M. and Buehler, M.J. (2012). Nonlinear material behaviour of spider silk yields robust webs. *Nature*, 482, 72-76.

Creager, M.S., Jenkins, J.E., Thagard-Yeaman, L.A., Brooks, A.E., Jones, J.A., Lewis, R.V., Holland, G.P. and Yarger, J.L. (2010). Solid-state NMR comparison of various spiders' dragline silk fiber. *Biomacromolecules*, 11, 2039-2043.

Crews, S.C. and Opell, B.D. (2006). The features of capture threads and orb-webs produced by unfed *Cyclosa turbinata* (Araneae: Araneidae). Journal of Arachnology, 34, 427-434.

Dawkins, R. (1982). The extended phenotype: The long reach of the gene Oxford: Oxford University Press.

Denny, M. (1976). The physical properties of spider's silk and their role in the design of orb-webs. *Journal of Experimental Biology,* 65, 483-506.

Dicko, C., Kenney, J.M., Knight, D. and Vollrath, F. (2004a). Transition to a beta-sheet-rich structure in spidroin in vitro: The effects of pH and cations. *Biochemistry*, 43, 14080-14087.

Dicko, C., Porter, D., Bond, J., Kenney, J.M. and Vollrath, F. (2007). Structural disorder in silk proteins reveals the emergence of elastomericity. *Biomacromolecules*, 9, 216-221.

Dicko, C., Vollrath, F. and Kenney, J.M. (2004b). Spider silk protein refolding is controlled by changing pH. *Biomacromolecules*, 5, 704-710.

Dong, Z., Lewis, R.V. and Middaugh, C.R. (1991). Molecular mechanisms of spider silk elasticity. *Archives of Biochemistry and Biophysics*, 284, 53-57.

Dyer, A.G. (2012). The mysterious cognitive abilities of bees: Why models of visual processing need to consider experience and individual differences in animal performance. *The Journal of Experimental Biology*, 215, 387-395.

Eberhard, W. (1969). Computer simulation of orb-web construction. *American Zoologist*, 9, 229-238.

Eberhard, W. (1986a). Effects of orb-web geometry on prey interception and retention. In: W. A. Shear Eds, Spiders: Webs, Behavior and Evolution. Stanford: Stanford University Press.

Eberhard, W. (1986b). Ontogentic changes in the web of *Epeirotypus* sp. (Araneae, Theridiomatidae). *Journal of Arachnology,* 14, 126-128.

Eberhard, W. (1987). Effects of gravity on temporary spiral construction by *Leucauge mariana* (Araneae: Araneidae). *Journal of Ethology*, 5, 29-36.

Eberhard, W.G. (1988). Behavioral flexibility in orb web construction : Effects of supplies in different silk glands and spider size and weight. *Journal of Arachnology*, 16, 295-302.

Eberhard, W.G., Agnarsson, I. and Levi, H.W. (2008). Web forms and the phylogeny of theridiid spiders (Araneae: Theridiidae): Chaos from order. *Systematics and Biodiversity*, 6, 1-61.

Eberhard, W.G. and Hesselberg, T. (2012). Cues that spiders (Araneae: Araneidae, Tetragnathidae) use to build orbs: Lapses in attention to one set of cues because of dissonance with others? *Ethology*, 118, 610-620.

Edmonds, D.T. and Vollrath, F. (1992). The contribution of atmospheric water vapour to the formation and efficiency of a spider's capture web. *Proceedings: Biological Sciences*, 248, 145-148.

Eisner, T., Alsop, R. and Ettershank, G. (1964). Adhesiveness of spider silk. *Science*, 146, 1058-1061.

Eles, P.T. and Michal, C.A. (2004). Strain dependent local phase transitions observed during controlled supercontraction reveal mechanisms in spider silk. *Macromolecules*, 37, 1342-1345.

Elices, M., Plaza, G.R., Pérez-Rigueiro, J. and Guinea, G.V. (2011). The hidden link between supercontraction and mechanical behavior of spider silks. *Journal of the Mechanical Behavior of Biomedical Materials*, 4, 658-669.

Ene, R., Papadopoulos, P. and Kremer, F. (2010). Partial deuteration probing structural changes in supercontracted spider silk. *Polymer*, 51, 4784-4789.

Foelix, R.F. (1987). Biology of Spiders (1st). Cambridge: Harvard University Press.

Frohlich, C. and Buskirk, R.E. (1982). Transmission and attenuation of vibration in orb spider webs. *Journal of Theoretical Biology*, 95, 13-36.

Gabriel, W. (2005). How stress selects for reversible phenotypic plasticity. Journal of Evolutionary *Biology*, 18, 873-883.

Gan, W., Liu, F., Zhang, Z. and Li, D. (2010). Predator perception of detritus and eggsac decorations spun by orb-web spiders *Cyclosa octotuberculata*: Do they function to camouflage the spiders? *Current Zoology*, 56,

Garrido, M.A., Elices, M., Viney, C. and Perez-Rigueiro, J. (2002a). The variability and interdependence of spider dragline tensile properties. *Polymer*, 43, 4495-4502.

Garrido, M.A., Elices, M., Viney, C. and Pérez-Rigueiro, J. (2002b). Active control of spider silk strength : Comparison of dragline spun on vertical and horizontal surfaces. *Polymer*, 43, 1537-1540.

Gosline, J.M., Denny, M.W. and DeMont, M.E. (1984). Spider silk as rubber. *Nature*, 309, 551-552.

Gosline, J.M., Guerette, P.A., Ortlepp, C.S. and Savage, K.N. (1999). The mechanical design of spider silks: From fibroin sequence to mechanical function. *Journal of Experimental Biology*, 202, 3295-3303.

Grubb, D.T. and Jelinski, L.W. (1997). Fiber morphology of spider silk: The effects of tensile deformation. *Macromolecules*, 30, 2680-2687.

Grubb, D.T. and Ji, G. (1999). Molecular chain orientation in supercontracted and re-extended spider silk. *International Journal of Biological Macromolecules*, 24, 203-210.

Guehrs, K.H., Schlott, B., Grosse, F. and Weisshart, K. (2008). Environmental conditions impinge on dragline silk protein composition. *Insect Molecular Biology*, 17, 553-564.

Guess, K.B. and Viney, C. (1998). Thermal analysis of major ampullate (drag line) spider silk: The effect of spinning rate on tensile modulus. *Thermochimica Acta*, 315, 61-66.

Guinea, G.V., Elices, M., Pérez-Rigueiro, J. and Plaza, G.R. (2005a). Stretching of supercontracted fibers: A link between spinning and the variability of spider silk. *Journal of Experimental Biology,* 208, 25-30.

Guinea, G.V., Elices, M., Real, J.I., Gutierrez, S. and Perez-Rigueiro, J. (2005b). Reproducibility of the tensile properties of spider (*Argiope trifasciata*) silk obtained by forced silking. *Journal of Experimental Zoology Part A-Comparative Experimental Biology*, 303A, 37-44.

Hansell, M. (2005). Animal Architecture Oxford: Oxford University Press.

Harmer, A.M.T. (2009). Elongated orb-webs of Australian ladder-web spiders (Araneidae: Telaprocera) and the significance of orb-web elongation. *Journal of Ethology*, 27, 453-460.

Harmer, A.M.T., Blackledge, T.A., Madin, J.S. and Herberstein, M.E. (2011). High-performance spider webs: Integrating biomechanics, ecology and behaviour. *Journal of the Royal Society Interface*, 8, 457-471.

Harmer, A.M.T. and Herberstein, M.E. (2009). Taking it to extremes: What drives extreme web elongation in Australian ladder web spiders (Araneidae: *Telaprocera maudae*)? *Animal Behaviour*, 78, 499-504.

Harmer, A.M.T. and Herberstein, M.E. (2010). Functional diversity of ladder-webs: Moth specialization or optimal area use? *Journal of Arachnology,* 38, 119-122.

Harwood, J.D. and Obrycki, J.J. (2005). Web-construction behavior of linyphiid spiders (Araneae, Linyphiidae): Competition and co-existence within a generalist predator guild. *Journal of Insect Behavior,* 18, 593-607.

Hawthorn, A.C. and Opell, B.D. (2002). Evolution of adhesive mechanisms in cribellar spider prey capture thread: Evidence for van der Waals and hygroscopic forces. *Biological Journal of the Linnean Society*, 77, 1-8.

Hayashi, C.Y. and Lewis, R.V. (1998). Evidence from flagelliform silk cDNA for the structural basis of elasticity and modular nature of spider silks. *Journal of Molecular Biology*, 275, 773-784.

Hayashi, C.Y., Shipley, N.H. and Lewis, R.V. (1999). Hypotheses that correlate the sequence, structure, and mechanical properties of spider silk proteins. *International Journal of Biological Macromolecules*, 24, 271-275.

Heiling, A.M. and Herberstein, M.E. (1999). The role of experience in web-building spiders (Araneidae). *Animal Cognition*, 2, 171-177.

Heiling, A.M. and Herberstein, M.E. (2000). Interpretations of orb-web variability: A review of past and current ideas. *Ekologia*, 19, 97-106.

Herberstein, M., Craig, C., Coddington, J.A. and Elgar, M. (2000a). The functional significance of silk decorations of orb-web spiders: A critical review of the empirical evidence. *Biological Reviews*, 75, 649-669.

Herberstein, M., Craig, C. and Elgar, M. (2000b). Foraging strategies and feeding regimes: Web and decoration investment in *Argiope keyserlingi* Karsch (Araneae: Araneidae). *Evolutionary Ecology Research*, 2, 69-80.

Herberstein, M.E., Gaskett, A.C., Glencross, D., Hart, S., Jaensch, S. and Elgar, M.A. (2000c). Does the presence of potential prey affect web design in *Argiope keyserlingi* (Araneae, Araneidae)? *Journal of Arachnology*, 28,

Herberstein, M.E. and Heiling, A. (1998). Does mesh height influence prey length in orb-web spiders (Araneae)? *European Journal of Entomology,* 95, 367-371.

Herberstein, M.E. and Tso, I.M. (2000). Evaluation of formulae to estimate the capture area and mesh height of orb webs (Araneoidea, Araneae). *Journal of Arachnology,* 28, 180-184.

Herberstein, M.E. and Tso, I.M. (2011). Spider webs: Evolution, diversity and plasticity. In: M. E. Herberstein Eds, Spider Behavior: Flexibility and Versatility. New York: Cambridge University Press.

Hesselberg, T. (2010). Ontogenetic changes in web design in two orb-web spiders. *Ethology*, 116, 535-545.

Hesselberg, T. and Vollrath, F. (2004). The effects of neurotoxins on web-geometry and web-building behaviour in *Araneus diadematus* Cl. *Physiology & behavior,* 82, 519-529.

Hesselberg, T. and Vollrath, F. (2006). Temperature affects both web spider response time and prey escape speed. *Bulletin of the British Arachnological Society*, 13, 275-280.

Higgins, L. and Rankin, M.A. (1999). Nutritional requirements for web synthesis in the tetragnathid spider *Nephila clavipes*. *Physiological Entomology*, 24, 263-270.

Higgins, L.E. (1992). Developmental plasticity and fecundity in the orb-weaving spider *Nephila clavipes*. *Journal of Arachnology*, 20, 94-106.

Higgins, L.E. (2006). Quantitative shifts in orb-web investment during development in *Nephila clavipes* (Araneae, Nephilidae). *Journal of Arachnology*, 34, 374-386.

Higgins, L.E. and Buskirk, R.E. (1992). A trap-building predator exhibits different tactics for different aspects of foraging behaviour. *Animal Behaviour*, 44, 485-499.

Higgins, L.E., Townley, M.A., Tillinghast, E.K. and Rankin, M.A. (2001). Variation in the chemical composition of orb webs built by the spider *Nephila clavipes* (Araneae, Tetragnathiidae). *Journal of Arachnology*, 29, 82-94.

Hinman, M.B., Jones, J.A. and Lewis, R.V. (2000). Synthetic spider silk: A modular fiber. *Trends in Biotechnology*, 18, 374-379.

Jelinski, L.W., Blye, A., Liivak, O., Michal, C., LaVerde, G., Seidel, A., Shah, N. and Yang, Z. (1999). Orientation, structure, wet-spinning, and molecular basis for supercontraction of spider dragline silk. *International Journal of Biological Macromolecules*, 24, 197-201.

Jenkins, J.E., Creager, M.S., Lewis, R.V., Holland, G.P. and Yarger, J.L. (2010). Quantitative correlation between the protein primary sequences and secondary structures in spider dragline silks. *Biomacromolecules*, 11, 192-200.

Jorger, K.M. and Eberhard, W. (2006). Web construction and modification by *Achaearanea tesselata* (Araeae, Theridiidae). *Journal of Arachnology,* 34,

Kishida, O., Trussell, G.C., Mougi, A. and Nishimura, K. (2010). Evolutionary ecology of inducible morphological plasticity in predator–prey interaction: Toward the practical links with population ecology. *Population ecology,* 52, 37-46.

Klarner, D. and Barth, F.G. (1982). Vibratory signals and prey capture in orb-weaving spiders (*Zygiella x-notata*, *Nephila clavipes*, Araneidae). *Journal of Comparative Physiology,* 148, 445-455.

Knight, D.P., Knight, M.M. and Vollrath, F. (2000). Beta transition and stress-induced phase separation in the spinning of spider dragline silk. *International Journal of Biological Macromolecules,* 27, 205-210.

Knight, D.P. and Vollrath, F. (1999). Liquid crystals and flow elongation in a spider's silk production line. *Proceedings of the Royal Society of London Series B-Biological Sciences*, 266, 519-523.

Knight, D.P. and Vollrath, F. (2001). Changes in element composition along the spinning duct in a *Nephila* spider. *Naturwissenschaften*, 88, 179-182.

Ko, F.K. and Jovicic, J. (2004). Modeling of mechanical properties and structural design of spider web. *Biomacromolecules*, 5, 780.

Kojić, N., Bico, J., Clasen, C. and McKinley, G.H. (2006). *Ex vivo* rheology of spider silk. *Journal of Experimental Biology*, 209, 4355-4362.

Krink, T. and Vollrath, F. (1997). Analysing spider web-building behaviour with rule-based simulations and genetic algorithms. *Journal of Theoretical Biology*, 185, 321-331.

Krink, T. and Vollrath, F. (1999). A virtual robot to model the use of regenerated legs in a web-building spider. *Animal Behaviour*, 57, 223-232.

Kümmerlen, J., van Beek, J.D., Vollrath, F. and Meier, B.H. (1996). Local structure in spider dragline silk investigated by two-dimensional spin-diffusion nuclear magnetic resonance. *Macromolecules*, 29, 2920-2928.

Landolfa, M.A. and Barth, F.G. (1996). Vibrations in the orb web of the spider *Nephila clavipes*: Cues for discrimination and orientation. *Journal of Comparative Physiology A-Sensory Neural And Behavioral Physiology,* 179, 493-508.

Levi, H.W. (1997). The American orb weavers of the genera *Mecynogea, Manogea, Kapogea* and *Cyrtophora* (Araneae: Araneidae). *Bulletin of the Museum of Comparative Zoology*, 155, 215-255.

Liao, C.-P., Chi, K.-J. and Tso, I.-M. (2009). The effects of wind on trap structural and material properties of a sit-and-wait predator. *Behavioral Ecology*, 20, 1194-1203.

Lin, L.H., Edmonds, D.T. and Vollrath, F. (1995). Structural engineering of an orb-spider's web. *Nature*, 373, 146-148.

Liu, Y., Shao, Z. and Vollrath, F. (2005). Extended wet-spinning can modify spider silk properties. *Chemical Communications*, 19, 2489-2491.

Liu, Y., Sponner, A., Porter, D. and Vollrath, F. (2008). Proline and processing of spider silks. *Biomacromolecules*, 9, 116-121.

Lubin, Y. and Henschel, J. (1996). The influence of food supply on foraging behaviour in a desert spider. *Oecologia*, 105, 64-73.

Lubin, Y., Kotzman, M. and Ellner, S. (1991). Ontogenetic and seasonal changes in webs and websites of a desert widow spider. *Journal of Arachnology*, 19, 40-48.

Madsen, B., Shao, Z. and Vollrath, F. (1999). Variability in the mechanical properties of spider silks on three levels: Interspecific, intraspecific and intraindividual. *International Journal of Biological Macromolecules*, 24, 301-306.

Madsen, B. and Vollrath, F. (1999). Mechanics and morphology of silk drawn from anesthetized spiders. *Naturwissenschaften*, 87, 148-153.

Manicom, C., Schwarzkopf, L., Alford, R.A. and Schoener, T.W. (2008). Self-made shelters protect spiders from predation. *Proceedings of the National Academy of Sciences,* 105, 14903-14907.

Masters, W.M. (1984). Vibrations in the orbwebs of *Nuctenea sclopetaria* (Araneidae) .1. Transmission through the web. *Behavioral Ecology and Sociobiology,* 15, 207-215.

Masters, W.M., Markl, H.S. and Moffat, A.J.M. (1986). Transmission of vibrations in a spider's web. In: W. A. Shear Eds, Spiders: Webs, Behavior and Evolution. Stanford: Stanford University Press.

Mayntz, D., Toft, S. and Vollrath, F. (2003). Effects of prey quality and availability on the life history of a trap-building predator. *Oikos*, 101, 631-638.

Mayntz, D., Toft, S. and Vollrath, F. (2009). Nutrient balance affects foraging behaviour of a trap-building predator. *Biology Letters,* 5, 735-738.

Miner, B.G., Sultan, S.E., Morgan, S.G., Padilla, D.K. and Relyea, R.A. (2005). Ecological consequences of phenotypic plasticity. *Trends in Ecology & Evolution*, 20, 685-692.

Miyashita, T. (1997). Factors affecting the difference in foraging success in three co-existing *Cyclosa* spiders. *Journal of Zoology,* 242, 137-149.

Mougi, A., Kishida, O. and Iwasa, Y. (2011). Coevolution of phenotypic plasticity in predator and prey: Why are inducible offenses rarer than inducible defenses? *Evolution*, 65, 1079-1087.

Nakata, K. (2007). Prey detection without successful capture affects spider's orb-web building behaviour. *Naturwissenschaften*, 94, 853-857.

Nakata, K. (2008). Spiders use airborne cues to respond to flying insect predators by building orb-web with fewer silk thread and larger silk decorations. *Ethology*, 114, 686-692.

Nakata, K. (2009). To be or not to be conspicuous: The effects of prey availability and predator risk on spider's web decoration building. *Animal Behaviour,* 78, 1255-1260.

Nakata, K. (2012). Plasticity in an extended phenotype and reversed up-down asymmetry of spider orb webs. *Animal Behaviour,*

Nakata, K. and Zschokke, S. (2010). Upside-down spiders build upside-down orb webs: Web asymmetry, spider orientation and running speed in *Cyclosa. Proceedings of the Royal Society B: Biological Sciences,* 277, 3019-3025.

Nentwig, W. (1987). The prey of spiders. In: W. Nentwig Ed, Ecophysiology of Spiders. Berlin: Springer-Verlag.

Nentwig, W. and Heimer, S. (1987). Ecological aspects of spider webs. In: W. Nentwig Eds, Ecophysiology of Spiders. Berlin: Springer-Verlag.

Ohgo, K., Kawase, T., Ashida, J. and Asakura, T. (2006). Solid-state NMR analysis of a peptide (Gly-Pro-Gly-Gly-Ala)$_6$-Gly derived from a flagelliform silk sequence of *Nephila clavipes. Biomacromolecules*, 7, 1210-1214.

Opell, B.D. (1994). The ability of spider cribellar prey capture thread to hold insects with different surface features. *Functional Ecology,* 8, 145-150.

Opell, B.D. (1995). Do static electric forces contribute to the stickiness of a spider's cribellar prey capture threads? *Journal of Experimental Zoology*, 273, 186-189.

Opell, B.D. (1997). The material cost and stickiness of capture threads and the evolution of orb-weaving spiders. *Biological Journal of the Linnean Society,* 62, 443-458.

Opell, B.D. (1999). Redesigning spider webs: Stickiness, capture area and the evolution of modern orb-webs. *Evolutionary Ecology Research,* 1, 503-516.

Opell, B.D. and Bond, J.E. (2001). Changes in the mechanical properties of capture threads and the evolution of modern orb-weaving spiders. *Evolutionary Ecology Research*, 3, 567-581.

Opell, B.D. and Hendricks, M.L. (2007). Adhesive recruitment by the viscous capture threads of araneoid orb-weaving spiders. *Journal of Experimental Biology,* 210, 553-560.

Opell, B.D., Karinshak, S.E. and Sigler, M.A. (2011). Humidity affects the extensibility of an orb-weaving spider's viscous thread droplets. *The Journal of Experimental Biology,* 214, 2988-2993.

Opell, B.D., Lipkey, G.K., Hendricks, M.L. and Vito, S.T. (2009). Daily and seasonal changes in the stickiness of viscous capture threads in *Argiope aurantia* and *Argiope trifasciata* orb-webs. *Journal of Experimental Zoology Part A-Ecological Genetics and Physiology*, 311A, 217-225.

Opell, B.D. and Schwend, H.S. (2007). The effect of insect surface features on the adhesion of viscous capture threads spun by orb-weaving spiders. *Journal of Experimental Biology*, 210, 2352-2360.

Ortlepp, C.S. and Gosline, J.M. (2004). Consequences of forced silking. *Biomacromolecules*, 5, 727-731.

Osaki, S. (1996). Spider silk as mechanical lifeline. *Nature*, 384, 419.

Osaki, S. (1999). Is the mechanical strength of spider's drag-lines reasonable as lifeline? *International Journal of Biological Macromolecules*, 24, 283-287.

Oxford, G. and Gillespie, R. (1998). Evolution and ecology of spider coloration. *Annual Review of Entomology,* 43, 619-643.

Pan, Z.J., Li, C.P. and Xu, Q. (2004). Active control on molecular conformations and tensile properties of spider silk. *Journal of Applied Polymer Science*, 92, 901-905.

Pan, Z.J. and Liu, M. (2009). Effects of drawing speed and water on microstructure and mechanical properties of artificially spun spider dragline silk. *Fibers and Polymers*, 10, 285-289.

Parkhe, A.D., Seeley, S.K., Gardner, K., Thompson, L. and Lewis, R.V. (1997). Structural studies of spider silk proteins in the fiber. *Journal of Molecular Recognition,* 10, 1-6.

Perez-Rigueiro, J., Elices, M., Plaza, G., Real, J.I. and Guinea, G.V. (2005). The effect of spinning forces on spider silk properties. *Journal of Experimental Biology*, 208, 2633-2639.

Pigliucci, M. (2005). Evolution of phenotypic plasticity: Where are we going now? *Trends in Ecology & Evolution*, 20, 481-486.

Pigliucci, M., Murren, C.J. and Schlichting, C.D. (2006). Phenotypic plasticity and evolution by genetic assimilation. *Journal of Experimental Biology,* 209, 2362-2367.

Prestwich, K. (1977). The energetics of web-building in spiders. *Comparative Biochemistry and Physiology Part A: Physiology,* 57, 321-326.

Pulz, R. (1987). Thermal and water relations. In: W. Nentwig Eds, Ecophysiology of Spiders. Berlin: Springer-Verlag.

Puthoff, J.B., Prowse, M.S., Wilkinson, M. and Autumn, K. (2010). Changes in materials properties explain the effects of humidity on gecko adhesion. *Journal of Experimental Biology*, 213, 3699-3704.

Rao, D. (2009). Experimental evidence for the amelioration of shadow competition in an orb-web spider through the 'ricochet' effect. *Ethology*, 115, 691-697.

Rao, D. (2010). Stingless bee interception is not affected by variations in spider silk decoration. *Journal of Arachnology*, 38, 157-161.

Rao, D., Cheng, K. and Herberstein, M.E. (2008). Stingless bee response to spider webs is dependent on the context of encounter. *Behavioral Ecology and Sociobiology*, 63, 209-216.

Rauscher, S., Baud, S., Miao, M., Keeley, Fred W. and Pomès, R. (2006). Proline and glycine control protein self-organization into elastomeric or amyloid fibrils. *Structure* (London, England : 1993), 14, 1667-1676.

Rising, A., Johansson, J., Larson, G., Bongcam-Rudloff, E., Engström, W., Hjälm, G. (2007). Major ampullate spidroins from *Euprosthenops australis*: Multiplicity at protein, mRNA and gene levels. *Insect Molecular Biology,* 16, 551-561.

Rising, A., Nimmervoll, H., Grip, S., Fernandez-Arias, A., Storckenfeldt, E., Knight, D.P., Vollrath, F. and Engström, W. (2005). Spider silk proteins - Mechanical property and gene sequence. *Zoological Science,* 22, 273-281.

Robinson, M.H. and Lubin, Y.D. (1979). Specialists and generalists: The ecology and behavior of some web-building spiders from Papua New Guinea. *Pacific Insects*, 21, 97-132.

Rodríguez S, R. (2000). Memory of captured prey in three web spiders (Araneae: Araneidae, Linyphiidae, Tetragnathidae). *Animal Cognition,* 3, 91-97.

Sahni, V., Blackledge, T.A. and Dhinojwala, A. (2011). Changes in the adhesive properties of spider aggregate glue during the evolution of cobwebs. *Scientific Reports*, 1, 41.

Sahni, V., Harris, J., Blackledge, T.A. and Dhinojwala, A. (2012). Cobweb-weaving spiders produce different attachment discs for locomotion and prey capture. *Nature Communications*, 3, 1106.

Salem, G.J., Zernicke, R.F. and Barnard, R.J. (1992). Diet-related changes in mechanical properties of rat vertebrae. American *Journal of Physiology - Regulatory, Integrative and Comparative Physiology,* 262, R318-R321.

Salomon, M. (2007). Western black widow spiders express state-dependent web-building strategies tailored to the presence of neighbours. *Animal Behaviour,* 73, 865.

Sandoval, C.P. (1994). Plasticity in web design in the spider *Parawixia bistriata*: A response to variable prey type. *Functional Ecology*, 8, 701-707.

Savage, K.N. and Gosline, J.M. (2008a). The effect of proline on the network structure of major ampullate silks as inferred from their mechanical and optical properties. *Journal of Experimental Biology*, 211, 1937-1947.

Savage, K.N. and Gosline, J.M. (2008b). The role of proline in the elastic mechanism of hydrated spider silks. *Journal of Experimental Biology*, 211, 1948-1957.

Scharf, I., Lubin, Y. and Ovadia, O. (2011). Foraging decisions and behavioural flexibility in trap-building predators: A review. *Biological Reviews,* 86, 626-639.

Scheibel, T. (2004). Spider silks: recombinant synthesis, assembly, spinning, and engineering of synthetic proteins. *Microbial Cell Factories*, 3, 14.

Schneider, J.M. and Vollrath, F. (1998). The effect of prey type on the geometry of the capture web of *Araneus diadematus*. *Naturwissenschaften*, 85, 391-394.

Seah, W.K. and Li, D. (2002). Stabilimentum variations in *Argiope versicolor* (Araneae: Araneidae) from Singapore. *Journal of Zoology*, 258, 531-540.

Sensenig, A., Agnarsson, I. and Blackledge, T.A. (2010). Behavioural and biomaterial coevolution in spider orb webs. *Journal of Evolutionary Biology,* 23, 1839-1856.

Sensenig, A.T., Lorentz, K.A., Kelly, S.P. and Blackledge, T.A. (2012). Spider orb webs rely on radial threads to absorb prey kinetic energy. *Journal of the Royal Society Interface*, 9, 1880-1891.

Sherman, P.M. (1994). The orb-web: An energetic and behavioural estimator of a spider's dynamic foraging and reproductive strategies. *Animal Behaviour*, 48, 19-34.

Simmons, A., Ray, E. and Jelinski, L.W. (1994). Solid-state ^{13}C NMR of *Nephila clavipes* dragline silk establishes structure and identity of crystalline regions. *Macromolecules*, 27, 5235-5237.

Swanson, B.O., Blackledge, T.A., Beltran, J. and Hayashi, C.Y. (2006). Variation in the material properties of spider dragline silk across species. *Applied Physics A-Materials Science & Processing,* 82, 213-218.

Tai, P.-L., Hwang, G.-Y. and Tso, I.-M. (2004). Inter-specific sequence conservation and intra-individual sequence variation in a spider silk gene. *International Journal of Biological Macromolecules*, 34, 237-243.

Tan, E.J. and Li, D. (2009). Detritus decorations of an orb-weaving spider, *Cyclosa mulmeinensis* (Thorell): For food or camouflage? *Journal of Experimental Biology*, 212, 1832-1839.

Tarakanova, A. and Buehler, M.J. (2012). The role of capture spiral silk properties in the diversification of orb webs. *Journal of the Royal Society Interface*, 9, 3240-3248.

Taylor, A.M., Bonser, R.H.C. and Farrent, J.W. (2004). The influence of hydration on the tensile and compressive properties of avian keratinous tissues. *Journal of Materials Science*, 39, 939-942.

Termonia, Y. (1994). Molecular modeling of spider silk elasticity. *Macromolecules*, 27, 7378-7381.

Théry, M. and Casas, J. (2009). The multiple disguises of spiders: Web colour and decorations, body colour and movement. *Philosophical Transactions of the Royal Society B: Biological Sciences*, 364, 471-480.

Tillinghast, E.K. and Townley, M.A. (1994). Silk glands of Araneid spiders. In: D.L. Kaplan, W.W. Adams, B. Farmer and C. Viney Eds, Silk Polymers: Materials Science and Biotechnology (ACS Symposium Series). Washington: American Chemical Society

Townley, M.A., Bernstein, D.T., Gallagher, K.S. and Tillinghast, E.K. (1991). Comparative study of orb web hygroscopicity and adhesive spiral composition in three Araneid spiders. *Journal of Experimental Zoology*, 259, 154-165.

Townley, M.A., Tillinghast, E.K. and Neefus, C.D. (2006). Changes in composition of spider orb web sticky droplets with starvation and web removal, and synthesis of sticky droplet compounds. *Journal of Experimental Biology,* 209, 1463-1486.

Tseng, L. and Tso, I. (2009). A risky defence by a spider using conspicuous decoys resembling itself in appearance. *Animal Behaviour,* 78, 425-431.

Tso, I.-M., Shu-Ya Chiang, S.-Y. and Blackledge, T.A. (2007). Does the giant wood spider *Nephila pilipes* respond to prey variation by altering web or silk properties? *Ethology*, 113, 324-333.

Tso, I.-M., Wu, H.-C. and Hwang, I.-R. (2005). Giant wood spider *Nephila pilipes* alters silk protein in response to prey variation. *Journal of Experimental Biology,* 208, 1053-1061.

Tso, I.M. (1999). Behavioral response of *Argiope trifasciata* to recent foraging gain: A manipulative study. *American Midland Naturalist*, 141, 238-246.

Tso, I.M. (2004). The effect of food and silk reserve manipulation on decoration-building of *Argiope aetheroides*. *Behaviour*, 141, 603-616.

Uetz, G.W. (1989). The "ricochet effect" and prey capture in colonial spiders. *Oecologia*, 81, 154-159.

Uhl, G. (2008). Size dependent occurrence of different types of web decorations and a barrier web in the tropical spider Argiope argentata (Fabricius 1775)(Araneae Araneidae). *Tropical Zoology*, 21, 97-108.

van Beek, J.D., Hess, S., Vollrath, F. and Meier, B.H. (2002). The molecular structure of spider dragline silk: Folding and orientation of the protein backbone. *Proceedings of the National Academy of Science USA*, 99, 10266-10271.

Vehoff, T., Glisovic, A., Schollmeyer, H., Zippelius, A. and Salditt, T. (2007). Mechanical properties of spider dragline silk: Humidity, hysteresis, and relaxation. *Biophysical Journal*, 93, 4425-4432.

Venner, S., Bel-Venner, M.-C., Pasquet, A. and Leborgne, R. (2003). Body-mass-dependent cost of web-building behavior in an orb weaving spider, *Zygiella x-notata*. *Naturwissenschaften*, 90, 269-272.

Venner, S. and Casas, J. (2005). Spider webs designed for rare but life-saving catches. *Proceedings of the Royal Society of London B,* 272, 1587-1592.

Venner, S., Pasquet, A. and Leborgne, R. (2000). Web-building behaviour in the orb-weaving spider *Zygiella x-notata*: Influence of experience. *Animal Behaviour,* 59, 603-611.

Vincent, J.F.V. (2002). Arthropod cuticle: A natural composite shell system Composites Part A: *Applied Science and Manufacturing*, 33, 1311-1315.

Vollrath, F. (1986). Gravity as an orientation guide during web-construction in the orb spider *Araneus diadematus* (Araneae, Araneidae). *Journal of Comparative Physiology A*, 159, 275-280.

Vollrath, F., Downes, M. and Krackow, S. (1997). Design variability in web geometry of an orb-weaving spider. *Physiology & behavior,* 62, 735-743.

Vollrath, F. and Knight, D.P. (2001). Liquid crystalline spinning of spider silk. *Nature*, 410, 541-548.

Vollrath, F. and Köhler, T. (1996). Mechanics of silk produced by loaded spiders. Proceedings: *Biological Sciences*, 263, 387-391.

Vollrath, F., Madsen, B. and Shao, Z.Z. (2001). The effect of spinning conditions on the mechanics of a spider's dragline silk. *Proceedings of the Royal Society of London Series B-Biological Sciences*, 268, 2339-2346.

Vollrath, F. and Mohren, W. (1985). Spiral geometry in the garden spider's orb web. *Naturwissenschaften*, 72, 666-667.

Walcott, C. (1963). The effect of the web on vibration sensitivity in the spider, *Achaeranea tepidariorum* (KOCH). *Journal of Experimental Biology*, 40, 595-611.

Walcott, C. (1969). A spider's vibration receptor: Its anatomy and physiology. *American Zoologist,* 9, 133-144.

Walter, A. and Elgar, M.A. (2011). Signals for damage control: Web decorations in *Argiope keyserlingi* (Araneae: Araneidae). *Behavioral Ecology and Sociobiology,* 165, 1909-1915.

Walter, A. and Elgar, M.A. (2012). The evolution of novel animal signals: Silk decorations as a model system. *Biological Reviews,* 87, 686-700.

Walter, A., Elgar, M.A., Bliss, P. and Moritz, R.F.A. (2008). Wrap attack activates web-decorating behavior in *Argiope* spiders. *Behavioral Ecology*, 19, 799-804.

Warwicker, J.O. (1960). Comparative studies of fibroins: II. The crystal structures of various fibroins. *Journal of Molecular Biology*, 2, 350-362.

Watanabe, T. (1999). The influence of energetic state on the form of stabilimentum built by *Octonoba sybotides* (Araneae, Uloboridae). *Ethology*, 105, 719-725.

Watanabe, T. (2000). Web tuning of an orb-web spider, *Octonoba sybotides*, regulates prey-catching behaviour. *Proceedings Of The Royal Society Of London Series B-Biological Sciences,* 267, 565-569.

Watanabe, T. (2001). Effects of web design on the prey capture efficiency of the uloborid spider Octonoba sybotides under abundant and limited prey conditions. *Zoological Science*, 18, 585-590.

Wirth, E. and Barth, F.G. (1992). Forces in the spider orb web. *Journal of Comparative Physiology A*, 171, 359-371.

Work, R.W. (1976). The force-elongation behavior of web fibers and silks forcibly obtained from orb-web-spinning spiders. *Textile Research Journal*, 46, 485–492.

Work, R.W. (1977). Dimensions, birefringences, and force-elongation behavior of major and minor ampullate silk fibers from orb-web-spinning spiders. The effects of wetting on these properties. *Textile Research Journal*, 47, 650–662.

Work, R.W. and Morosoff, N. (1982). A physico-chemical study of the supercontraction of spider major ampullate silk fibers. *Textile Research Journal,* 52, 349-356.

Xu, M. and Lewis, R.V. (1990). Structure of a protein superfiber: Spider dragline silk. *Proceedings of the National Academy of Science USA*, 87, 7120-7124.

Yang, Z.T., Liivak, O., Seidel, A., LaVerde, G., Zax, D.B. and Jelinski, L.W. (2000). Supercontraction and backbone dynamics in spider silk: C-13 and

H-2 NMR studies. *Journal of the American Chemical Society*, 122, 9019-9025.

Zax, D.B., Armanios, D.E., Horak, S., Malowniak, C. and Yang, Z. (2004). Variation of mechanical properties with amino acid content in the silk of *Nephila clavipes*. *Biomacromolecules*, 5, 732-738.

Zevenbergen, J.M., Schneider, N.K. and Blackledge, T.A. (2008). Fine dining or fortress? Functional shifts in spider web architecture by the western black widow *Latrodectus hesperus*. *Animal Behaviour*, 76, 823-829.

Zschokke, S. and Nakata, K. (2010). Spider orientation and hub position in orb webs. *Naturwissenschaften*, 97, 43-52.

In: Spiders
Editor: Marq Santerre

ISBN: 978-1-62618-517-3

Chapter 2

SPIDER WEBS AS EXTENDED PHENOTYPES

Sean J. Blamires*
Department of Life Science, Tunghai University, Taichung, Taiwan

ABSTRACT

An animal's structure is an expression of its genotype beyond the confines of the body, so is considered an extended phenotype. The fitness consequences of extended phenotypes are not well defined because extended phenotypes affect fitness via interactions with both the environment and the somatic phenotype and these interactions are not well understood. Here I illustrate that web-building spiders make ideal models for testing hypotheses regarding the functional significance and adaptive benefits of extended phenotypes. A review of the literature reveals that spider webs, silks and additional components, e.g. decorations, decoys and barrier webs, represent expressions of (i) a multitude of behaviours including foraging and courtship, (ii) the spider's ontogeny, (ii) the sensory capacities of the spider and/or its predators and prey, and (iv) the phenotypic expression of any infecting parasitoids. I herein assess the adaptive benefits of expressing webs as extended phenotypes and outline the feedback cycles that influence spider fitness. Niche construction is an under explored aspect of extended phenotypes. I thus describe how it might play a role in enhancing spider fitness. I will lastly outline some research applications of using the 'web as an extended phenotype' model, in investigations of: (i) invasive and conservation

* E-mail address: sblamires@thu.edu.tw.

biology of spiders, (ii) spider phylogenetics, and (iii) using webs for biomonitoring, among others.

EXTENDED PHENOTYPES: AN INTRODUCTION

An extended phenotype is a product of an animal's genotype expressed externally [1-3]. Animal structures, e.g. beaver dams, termite mounds, bird and reptile nests, burrows and spider and antlion traps, thus are extended phenotypes [3-7]. Animal structures are extended phenotypes that serve two functions; creating a home or catching food [4,6,8-12]. Variations in expression is found in many extended phenotypes [4,5,11,13,14]. For instance, the architecture of spider and antlion traps vary with biotic (e.g. temperature, humidity and wind) and abiotic (e.g. prey types and predators) environmental features [6,14]. Such structural variation affects and/or feeds back on other phenotypic levels via niche construction (the ability of animals to modify the environment they experience) [8,14-16], ecosystem engineering (the ability of animals to modify the ecosystem they live in) [4,17], or other complex interactive mechanisms [6,13,18,19].

SPIDER WEBS AS MODEL EXTENDED PHENOTYPES

Web-building spiders make a good model for testing hypotheses about the heritability and adaptive benefits of extended phenotypes because: (i) the web is both a home and a prey catching trap, (ii) they are found across a wide range of habitats, (iii) they have a range of life histories, e.g. there are an abundance of generalists and specialists and diurnal, nocturnal and crepuscular species, and (iv) the web's architecture and the physical and chemical properties of the silk vary in response to changes in the biotic, e.g. prey types/nutrients, and abiotic, e.g. temperature, wind, environment [12,20-25].

1. The Web as a Depiction of Spider Behaviour

Foraging behaviours, including foraging plasticity, are well documented in web-building spiders, especially orb-weaving spiders (see [14]). This is because the web is the manifestation of the foraging strategy that can be easily measured [5,12,20-24,26-28]. Moreover, specific behaviours during web construction result in specific web architectures being built. These behaviours are predictable to the extent that computer simulations with specific input rules and algorithms can accurately predict the architectures of webs that spiders can produce in specific instances [29].

Modelers of spider foraging behaviours assume that variations in foraging behaviour are depicted by variations in web architecture. These are generally measured as the mean amount of variation in components such as web capture surface area, the width of the spaces between the sticky spirals (mesh size), the number of radii that traverse the spirals and the length or pattern of any decorations (stabilimenta); the conspicuous silken or inert structures added to the centre of the web by some spiders [30]. Spider orb webs are constructed from up to seven types of silk, each of which is secreted from a different gland and contributes specifically to different components of web architecture [31,32]. The radii are composed of silks produced by the major (MA) and minor ampullate (MiA) glands. The spiral silks on the other hand are derived from the aggregate and flagelliform glands, and silks derived from the aciniform gland are used in decorations [32,33].

The amino acid composition differs for each of these silks, for instance, MA and MiA silks are principally comprised (~85%) of short chain, readily synthesizable amino acids such as glycine and alanine, while aggregate and aciniform silks are composed of around 60% synthesizable amino acids. The metabolic costs of investment in each of the web components accordingly differs [31,34]. Since the different amino acids have different metabolic cost of synthesis [34], these variations may explain why the components are differentially expressed when compositions of prey captured or prey nutritional values vary [5,6,21,23,28,35]. Unfortunately, these costs are rarely included in models of web building spider foraging (but see [35]).

Additional components that may be added to orb webs include decorations made from silk, detritus, prey or eggsacs [18,19,30,36], or silk or inert barriers, decoys and retreats [19,37-39]. Aggregated webs may perform functions similar to barrier structures, i.e. as protection for the spider or to assistance in prey capture via "ricochet effects" [18], so may also be considered a kind of additional web component, but I will not elaborated on these herein.

The use and design of silken cruciform or linear web decorations is variable among and within individual *Agiope* spp. [5,21-23,40], for whom decorations are thought to lure prey by reflecting UV and/or blue light [30,36]. The presence of predators and prey, variations in nutritional intake and human/animal induced web damage have been suggested as factors inducing variability in web decoration designs in *Argiope* spp., and other orb-weaving spiders [5,18,19,21-23,40-42]. The three-dimensional barrier structures added to many two-dimensional webs (e.g. the semi-permanent barrier webs of some *Nephila* spp. and the permanent barrier webs of *Cyrtophora* spp.) provide protection from predators and may assist in prey capture efficiency [18,19,43,44]. Decorations and barrier webs, therefore, may be considered extended phenotypes depicting either, or a combination of, the spider's foraging, anti-predator and web protecting strategies.

Foraging is not the only behaviour that a spider's web and its decorations or other components depict. Other behaviours depicted include web site selection and anti-predator, reproductive and social behaviours [22,26,45-50]. Thus, measuring web architecture to estimate foraging investment may be misleading in assessments of the web's function and/or any associated costs and benefits to web building [6,14,31].

2. The Web as a Depiction of Reproductive Condition

The size and shape of many spider's webs varies with reproductive condition and strategy [26,51,52]. Web architectures, therefore, may also be considered extended phenotypes of reproductive condition. No study has as yet empirically determined the causal mechanisms, or indeed if there are causal mechanisms, associated with web architectural variations and reproductive condition. A reason for variations in architecture with reproductive condition may be because both webs and eggs cost the spider energy and nutrients, so web investment may be traded-off against egg development. Alternatively, in orb-weavers at least, reproductive females are often considerably large and specific web architectures are needed to withhold their greater mass [53-55]. A further explanation could be that web architectures vary with spider senescence [56] co-incidentally to reproductive state.

3. The Web as a Depiction of Ontogeny

The web architectures of spiders varies though ontogeny [57-59]. Accordingly, web architecture may be considered an extended phenotype depicting ontogeny. Allometric variations in silk production, the capacity to build webs, or to capture prey within webs, explain the variations in webs as spiders grow [59]. One aspect of web architecture that depicts strong ontogenetic variations in orb webs is the position of the hub relative to the centre of the orb, so called "up-down asymmetry" [59-63]. The smaller the spider the more centrally placed the hub is [59]. Second to third instar orb web spiders usually build orb webs with a central hub, while mature spiders build orbs with the hub much closer to the top of the web than the bottom [64,65]. The largest orb web spiders, e.g. *Nephila* spp., build webs with the most asymmetrically placed hubs [57,59]. This phenomenon suggests that it is the contrast in the ability of the spider to run upwards against gravity compared to downwards with gravity that causes the ontogenetic shifts in hub position; large spiders run up more laboriously than they do down [61].

In addition to hub position, the radii to spiral ratio, or the use or density of the barrier web, the use and/or design of web decorations, and the thickness and mechanical performance of web silks varies, depending on the species, within individual spiders through ontogeny [55,57,65-69]. In some species these ontogentic changes are thought to reflect its phylogenetic history, the so-called "biogenic law" [65,70,71]. However, in other species this has proven to be not the case [58,59,72]. In any event, a combination of web architecture, silk properties and use of additional components are good indicators of ontogeny in many spiders.

4. The Web as Depiction of Physiological Stress

Extreme environments (e.g. exceptionally high or low temperatures, humidity or strong winds) influences web architecture as well as the physical and chemical properties of the silks [20,24,25,73-75]. Moreover, the thickness of the radial, frame and dragline silks increases with ambient temperature and this may be a product of the spider drawing the silks at different speeds [73,76]. Starved spiders build webs with fewer radii but have a greater investment in spiral silks than do well fed spiders [77-79]. The spirals are nonetheless less sticky than are those of the well fed spiders [80-82]. Recently, nutritional stress has also been shown to affect web architectures. Spiders feeding on more nutritious prey or particular prey reared on nutrient-enhanced media produce more radii, spirals, and/or add longer decorations to their webs [5,23,27,28,35]. The webs of spiders fed the more nutritious prey are also stiffer and stickier [12]. Silk physical and chemical properties also vary with nutritional state of the spider [35,83]. Web architecture is thus an extended phenotype depicting the spider's energetic and/or nutritional state [5].

Physiological stresses, including stresses induced during molting and stresses associated with thermal and water balance, are among the multitude of purposes ascribed to web decorations [30,84-86]. Decorations and their designs and patterns may thus represent extended phenotypes of spider thermo-hydric condition among other physiological processes. Furthermore, there is evidence that the spectral properties and shape forms of many web decorations, such as those used by *Argiope* spp. and *Cyclosa* spp., are attuned to the visual sensitivities of predatory and/or prey insects [19,39,87-90], and variations in decoration form within individuals may be a response to predator and/or prey learning to associate the decorations with the spider [40,48,89-92]. Accordingly, decoration form and properties may reflect, and be an extended phenotype of, the visual and learning capabilities of the insects that encounter them.

5. The Web as Extended Sensory System

As previously stated, specific behaviours are used in specific sequences to build a web. These behaviours are most well depicted in the orb-weaving spiders and there appears to be homology among families [57,93,94], with potential exceptions among species exposed to extreme environments [95,96]. Nonetheless, the sequence of behaviours during web building is generally dependent on the spider receiving certain cues from the environment, such as the direction of gravity and air temperature [12,97-99], and from the web itself, such as the tactile stimuli generated in specific threads [100]. The specific behavioural response(s) to the cues may be dependent on spider weight and/or leg length [101]. The specific cues used may be predictors of the type of prey available and/or are good web performance indicators in a given environment [12,99,102]. The ability of spiders to detect cues or to respond to a multitude of cues, thus determines the architecture of the web that is built [99]. Web architecture thus is a depiction of the spider's ability to detect and act on specific cues.

A spider additionally uses the web itself to detect changes in its environment and accordingly adjust web architecture. Spiders use vibrations propagated through the radii to detect the presence of prey [100]. The number of radii and the number of spiral threads that traverse each radii must be balanced so that the vibratory signal received from struggling prey in the web periphery is adequate to accurately locate the prey [100-102]. Research has found that artificially stimulating the web or the addition or removal of radii and spiral threads induces adjustments in web architecture [12,103,104]. Hence, specific web architectures feed back on and affect web building behaviours.

6. The Web as an Extended Phenotype of Parasitoids

Parasite and parasitoid manipulation of host behaviours is an additional example of an extended phenotype [1,3,105]. Upon deposition into a host body, a developing parasitoids larvae induces its host to forage for nutrients for its benefit until the parasitoid emerges and the host is induced to commit "suicide" [1,105-107]. A similar phenomenon has been described for when an ichneumonid wasp endoparasitizes an orb web spider. In some instances the spider my be induced to spin a cocoon with a durable web support before self-sacrifice so that the wasp larvae can continue to develop [108]. In other instances parasitoids may induces spiders to build dense webs in order to over winter before inducing the spider to commit "suicide" upon emergence [109]. Spiders suffer attacks from a range of parasitoids [108-115]. However, there have been few direct observations of parasitoid manipulation of spiders and the resultant webs built. It would be intriguing to know whether parisitoids can induce spiders to build webs to facilitate particular foraging outcomes, as they do to ants and other insects [105,116,117].

ADAPTIVE BENEFITS OF SPIDER WEB EXTENDED PHENOTYPES

1. Feedback Cycles

Figure 1 summarizes the feedback cycles associated with the spider web extended phenotype between (i) the environment, (ii) the somatic phenotype, (iii) the extended phenotypes(s), and (iv) fitness effects on the spider. The figure is a simplified multiple feedback system; in reality it is more complicated. For instance, the properties of spider silks, webs and additional components are inter-correlated [5,118,119] and/or co-vary across environments [20,21,24-27], so fitness effects of their plasticity across levels are also likely to co-vary. The principal fitness benefit of web (including all additional components) and silk plasticity seems to be to bestow spiders with the ability to cater the performance of the web to capture the most profitable or abundant prey available in a specified environment without undue costs [12,52,119].

Nevertheless, ontogeny, courtship, the presence of predators, conspecifics, allospecifics and parasitoids induces further variations in the design of webs, the properties of silks and influence the use of additional components [39,48,57,59, 66,96,104,108]. This then feeds back on fitness and spider physiology; and the environment via niche construction. Additionally, a multitude of interactive environmental cues induce variations in web building behaviours [99], influencing web architectures and, consequently, the structure and function of additional components thereby feeding back on spider physiology and fitness (Figure 1).

2. Niche Construction

Niche construction is the phenomena of animals altering key factors in their immediate environment to enhance their own probability for survival [120]. Animals that build structures thus are generally niche constructors [4]. There is some controversy surrounding niche construction with some authors suggesting it is not a process, but a by-product or specific component of the extended phenotype [121].

A niche constructing mechanism of web building by spiders and spider web plasticity has been proposed but remains to be empirically tested. Spiders use web architecture and additional components to increase the probability of capturing specific prey [19,44,53,57,90]. This means that spider webs may manipulate the nutritional environment for the benefit of the spider. It seems, given the complicity of spiders to build webs under laboratory conditions, the perfect system to test the extent to which niche construction adds to, feeds back on, or limits extended phenotypes. It has been postulated that spiders use webs to maximize their growth and/or fecundity niches while reducing predation by buffering against environmental perturbations [4,122,123]. This is congruent with optimal foraging and optimal performance explanative models of spider web plasticity [14], which suggest that spiders build webs that vary in architecture so as to most effectively capture the largest or most nutrient rich prey available in their ecosystem [12,78,124].

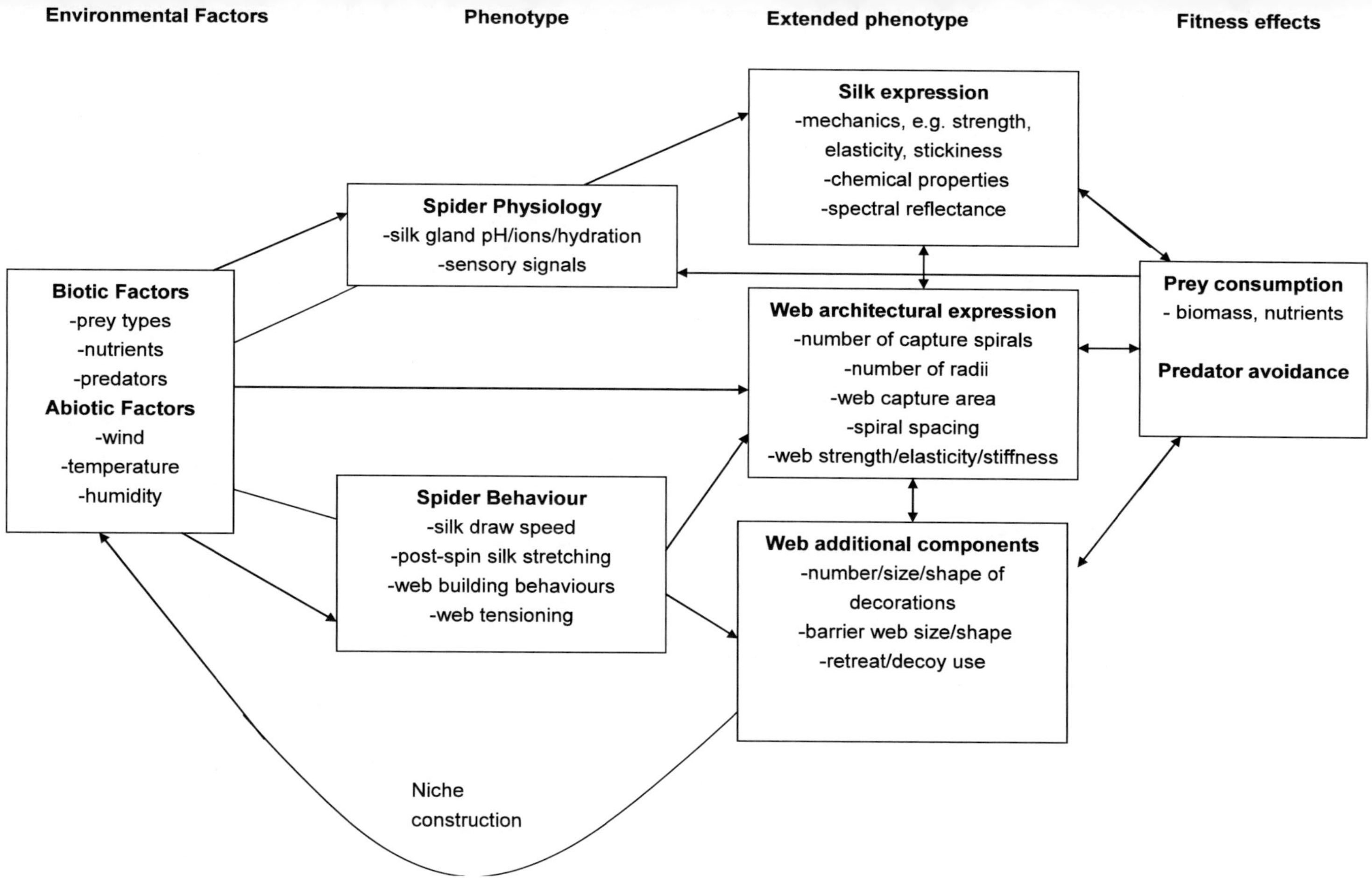

Figure 1. The feedback model of interactions between webs and (i) environmental factors, (ii) the phenotype, (iii) the extended phenotypes (webs, silk and additional components), and (iv) spider fitness, highlighting the role of niche construction as a feedback mechanism between the webs and the environment.

Research Applications

I outline three research applications of using the 'webs as extended phenotype' model: (i) invasive and conservation biology of spiders, (ii) phylogenetics and (iii) biomonitoring using spider webs. This is by no means an exhaustive list of the possible research applications. Additional potential applications may include using webs as a means to understand the complexities of phenotypic plasticity (see [3,5,14]), to assess spatial and temporal coevolutionary mosaics [125] between different spiders, spiders and their prey, or spiders and predators or parasitoids, and to assess the ecological drivers of sociality [49,126].

1. Invasive and Conservation Biology of Spiders

Extended phenotypes that are variable across environments provides the flexibility required for animals to expand the "fundamental niche", which is defined as the range of conditions that can be occupied in the absence of confounding influences such as predators or competitors [127,128].

Hence, animals that exhibit a multitude of traits, including the ability to construct extended phenotypes, are more likely to invade, survive and evolve in new environments than animals that do not.

Furthermore projected climate changes scenarios predict that particular animals will undergo range expansions while others will undergo range contractions [127,129].

Since animals with extended phenotypes may be protected from the mitigating influences of climate variations via niche construction [3,4,122,123] their ranges should advance in a changed environment. Indeed large numbers of exotic and native web building spider are identified as invasive across Asia, Europe and North America [130-133]. These invasions are of concern because of the potential negative impacts on the conservation of native insects and on agricultural processes since spiders are an apex predator in invertebrate ecosystems [134].

Accordingly, it is important to understand the facilitating factors of their range expansion, including the role of their extended phenotypes in facilitating adaptive plasticity [14] and niche construction in changed environments.

2. Phylogenetics

Phylogenetic studies of extended phenotypes may assist studies into the evolution and co-evolution of key phenotypic and behavioural innovations [4,135,136]. Indeed, phylogenetic analyses of spider web designs have been instrumental in enabling the identification of factors, such as predator diversification, predator and prey visual perceptibilities, and micro-environments, as drivers of the spider evolution and the diversification of web architectures [43,58,90,136,137]. While there remains controversy pertaining to the conclusions of these studies and whether or not evolution has acted directly on the structures, or the structures represent an "end-product" of a complex suite of physiological and behavioural traits that selection acts on [5,138], the work on spider webs serves as a good template for the examination of the selection pressure and constraints acting on the diversification of animals and their structures. An understanding of the selection pressures and constraints acting on structural variation may also be useful in explicating the evolutionary formation of high performance materials, such as spider silk (see [55,69,139-141] for examples).

3. Spider Webs and Silks as Bioindicators

Directly measuring pollutants, pesticides or heavy metals may provide information about the viability of ecosystems but is time consuming and potentially hazardous. An assessment of organisms whose abundance, morphology, behaviour or physiology can reliably indicate environmental integrity, so called "bioindicators", may be preferred because they provide information about the biological impact of the pollutants. To be a good bioindicator an organisms must: (1) provide reliable information, hence the population must be adequately sensitive across well-defined environmental variations, (2) respond to environmental change in proportion to the degree of change incurred, (3) be abundant, common and in a stable population, (4) have a well-understood biology and (5) be well-recognized [142].

Spiders are examples of animals used as bioindicators for soil productivity and atmospheric pesticides and heavy metal pollution [143-146]. Using spiders as bioindicators is, nonetheless, generally problematic because they move around and so are difficult to repeatedly sample at a single location. As the web depicts the behavioural, physiological and/or biochemical status of the spider and quantitatively reflect the population level responses to spatial and temporal environmental variation, webs may qualify as bioindicators in certain circumstances [10,144,147]. The properties and chemistry of silk are directly affected by the uptake of chemicals from the atmosphere or food [10,28,81,83,148-151] so spider silk may also qualify as a bioindicator.

ACKNOWLEDGMENT

This review was written while receiving a National Science Council of Taiwan postdoctoral fellowship (NSC-98-2811-B-029-002).

REFERENCES

[1] Dawkins, R., *The extended phenotype: the long reach of the gene*. 1982, Oxford: Oxford University Press.

[2] Schaedelin, F. C., and M. Taborsky, Extended phenotypes as signals. *Biol Rev*, 2009. 84: p. 293-313.

[3] Bailey, N. W., Evolutionary models of extended phenotypes. *Trends Ecol Evol*, 2012. 27: p. 561-569.

[4] Hansell, M., *Animal architecture*. 2005, Oxford: University of Oxford Press.

[5] Blamires, S. J., Plasticity in extended phenotypes: orb web architectural responses to variations in prey parameters. *J Exp Biol*, 2010. 213: p. 3207-3212.

[6] Scharf, I., Y. Lubin, and O. Ovadia, Foraging decisions and behavioural flexibility in trap-building predators: a review. *Biol Rev*, 2011. 86: p. 626-639.

[7] Sergio, F., J. Blas, G. Blanco, A. Tanferna, L. Lopez, A. Lemus, and F. Hiraldo, Raptor nest decorations are a reliable threat against conspecifics. *Science*, 2011. 331: p. 327-330.

[8] Whitham, T. G., E. Lonsdorf, J. A. Schweitzer, J. K. Bailey, D. G. Fischer, S. M. Shuster, R. L. Lindroth, S. C. Hart, G. J. Allen, C. A. Gehring, P. Keim, B. M. Potts, B. J. Rehill, S. P. Di Frazio, C. J. Le Roy, G. M. Wimp, and S. woolbright, "All effects of a gene on the world": extended phenotypes, feedbacks, and multi-level selection. *Ecoscience*, 2005. 12: p. 5-7.

[9] Downing, H. A., and R. L. Jeanne, A comparison of nest construction behavior in two species of *Polistes* paper wasps (Insecta, Hymenoptera: Vespidae). *J Ethol*, 1987. 5: p. 53-66.

[10] Tessier, L., J. L. Boisvert, L. B. M. Vought, and J. O. Lacoursiere, Anomalies on capture nets of *Hydropsyche slossonae* larvae (Trichoptera; Hydropsychidae), a potential indicator of chronic toxicity of malathion (organophosphate insecticide). *Aquat Toxicol*, 2000. 50: p. 125-139.

[11] Rushbrook, B. J., N. J. Dingemanse, and I. Barber, Repeatability in nest construction by male three-spined sticklebacks. *Anim Behav*, 2008. 75: p. 547-553.

[12] Blamires, S. J., I. C. Chao, C. P. Liao, and I. M. Tso, Multiple prey cues induce foraging flexibility in a trap-building predator. *Animal Behav*, 2011. 81: p. 955-961.

[13] Rushbrook, B. J., and I. Barber, A comparison of nest building by three-spined sticklebacks *Gasterosteus aculeatus* from still and flowing waters. *J Fish Biol*, 2008. 73: p. 746-752.

[14] Boutry, C., and S. J. Blamires, Plasticity in web architecture and silk mechanics. *Spiders: morphology, behavior and geographic distribution*. 2013, Nova: New York.

[15] Odling-Smee, J., Niche constructing phenotypes. *Role of behavior in evolution*. ed. H. C. Plotkin. 1988, Cambridge MA: MIT Press. p. 73-132.

[16] Miner, B. G., S. E. Sultan, S. G. Morgan, D. K. Padilla, and R. A. Relyea, Ecological consequences of phenotypic plasticity. *Trends Ecol Evol*, 2005. 20: p. 685-692.

[17] Jones, C. G., J. H. Lawton, and M. Scachak, Organisms as ecosystem engineers. *Oikos*, 1994. 69: p. 373-386.

[18] Blamires, S. J., Y. H. Lee , C. M. Chang, I. T. Lin, J. A. Chen, T. Y. Lin, and I. M. Tso, Multiple structures interactively influence prey capture efficiency in spider orb webs. *Anim Behav*, 2010. 80: p. 947-953.

[19] Tseng, H. J., R. C. Cheng, S. H. Wu, S. J. Blamires, and I. M. Tso, Trap barricading and decorating by a well-armored sit-and-wait predator: extra protection or prey attraction? *Behav Ecol Sociobiol*, 2011. 65: p. 2351-2359.

[20] Vollrath, F., M. Downes, and S. Krakow, Design variability in web geometry of an orb-weaving spider. *Physiol Behav*, 1997. 62: p. 735-743.

[21] Blamires, S. J., D. F. Hochuli, and M. B. Thompson, Does decoration building influence antipredator responses in an orb-web spider (*Argiope keyserlingi*) in its natural habitat? *Aust J Zool*, 2007. 55: p. 1-7.

[22] Blamires, S. J., M. B. Thompson, and D. F. Hochuli, Habitat selection and web plasticity by the orb spider *Argiope keyserlingi*: do they compromise foraging success for predator avoidance? *Austral Ecol*, 2007. 32: p. 551-563.

[23] Blamires, S. J., D. F. Hochuli, and M. B. Thompson, Prey protein influences growth and decoration building in the orb spider *Argiope keyserlingi*. *Ecol Entomol*, 2009.34: p. 545-550.

[24] Liao, C. P., K. J. Chi, and I. M. Tso, The effects of wind on trap structural and material properties of a sit-and-wait predator. *Behav Ecol* 2009. 20: p. 1194-1203.

[25] Wu, C. C., S. J. Blamires, C. L.Wu, and I. M. Tso, Wind-induces variations in spider web geometry and spiral droplet volume but not spiral stickiness. *J Exp Biol*, in press.

[26] Blackledge, T. A., and J. M. Zevenberg, Condition-dependent spider web architecture in the western black widow, *Latrodectus hesperus*. *Anim Behav*, 2007. 73: p. 855-864.

[27] Tso, I. M., S. Y. Chiang, and T. A. Blackledge, Does the giant wood spider *Nephila pilipes* respond to prey variation by altering web or silk properties? *Ethology*, 2007.113: p. 324-333.

[28] Mayntz D, S. Toft, and F. Vollrath, Nutrient balance affects foraging behaviour of a trap-building predator. *Biol Lett*, 2009. 5: p. 735-738.

[29] Krink, T., and F. Vollrath, Analysing spider web-building behaviour with rule-based simulations and genetic algorithms. *J Theor Biol*, 1997. 185: p. 321-331.

[30] Herberstein, M. E., C. L. Craig, J. A. Coddington, and Elgar, M. A. The functional significance of silk decorations of orb-web spiders: a critical review of the empirical evidence. *Biol Rev*, 2000. 75: p. 649-669.

[31] Craig, C. L., *Spiderwebs and silk: tracing evolution from molecules to genes to phenotypes*. 2003, Oxford: Oxford University Press.

[32] Vollrath, F., and D. P. Knight, Liquid crystalline spinning of spider silk. *Nature*, 2001. 410: p. 541-548.

[33] Blackledge, T. A., and C. Y. Hayashi, Silken toolkits: biomechanics of silk fibers spun by the orb web spider *Argiope argentata* (Fabricius 1775). *J Exp Biol*, 2006. 209: p. 2452-2461.

[34] Craig, C. L., M. Hsu, D. L. Kaplan, and M. E. Pierce, A comparison of the composition of silk proteins produced by spiders and insects. *Inter J Biol Macromol*, 1999. 24: p. 109-118.

[35] Blamires, S. J., and I. M. Tso, Nutrient mediated architectural plasticity of a predatory trap. *PLoS ONE*, 2013. 8: 54558.

[36] Thery, M., and J. Casas, The multiple disguises of spiders: web colour and decorations, body colour and movement. *Philos Trans Roy Soc B*, 2009. 364: p. 471-480.

[37] Bjorkman-Chiswell, B. T., M. M. Kulinski, R. L. Muscat, K. A. Nguyen, B. A. Norton, M. R. E. Symonds, and M. A. Elgar, Web-building spiders attract prey by storing decaying matter. *Naturwissenschaften*, 2004. 91: p. 245-248.

[38] Manicom, C., L. Schwarzkopf, R. A. Alford, and T. W. Schoener, Self-made shelters protect spiders from predation. *Proc Nat Acad Sci USA*, 2008. 105: p. 14903-14907.

[39] Tseng, L., and I. M. Tso, A risky defence by a spider using conspicuous decoys resembling itself in appearance. *Anim Behav*, 2009. 78: p. 425-431.

[40] Craig, C. L., S. G. Wolf, J. L. Davis, M. E. Hauber, and J. L. Maas, Signal polymorphism in the web-decorating spider *Argiope argentata* is correlated with reduced survivorship and the presence of stingless bees, its primary prey. *Evolution*, 2001. 55: p. 986-993.

[41] Watanabe, T., The influence of energetic state on the form of stabilimentum built by *Octonoba sybotides* (Araneae: Uloboridae). *Ethology*, 1999. 105: p. 719-725.

[42] Walter A., and M. A. Elgar, The evolution of novel animal signals: silk decorations as a model system. *Biol Rev*, 2012. 87: p. 686-700.

[43] Blackledge, T. A., J. A. Coddington, and R. G. Gillespie, Are three-dimensional spider webs defensive adaptations? *Ecol Lett*, 2003. 6: p. 13-18.

[44] Blamires, S .J., C. Hou, L. F. Chen, C. P. Liao, and I. M. Tso, Three-dimensional barricading of a predatory trap reduces predation and enhances prey capture. *Behav Ecol Sociobiol*, 2013. 67: in press.

[45] Tietjen, W. J., Effects of colony size on web structure and behavior of the social spider *Mallos gregalis* (Araneae, Dictynidae). *J Arachnol*, 1986.14: p. 145-157.

[46] Schoener, T. W., and D. A. Spiller, Stabilimenta characteristics of the spider *Argiope argentata* on small islands: support for the predator defense hypothesis. *Behav Ecol Sociobiol*, 1992. 31: p. 309–318.

[47] Blackledge, T. A., and J. W. Wenzel, State-determinate foraging decisions and web architecture in the spider *Dictyna volucripes* (Araneae Dictynidae). *Ethol Ecol Evol*, 2001. 13: p. 105-113.

[48] Li, D., and W. S. Lee, Predator-induced plasticity in web-building behaviour. *Anim Behav*, 2004. 67: p. 309-318.

[49] Yip, E., K. S. Powers, and L. Aviles, Cooperative capture of large prey solved scaling challenge faced by spider societies. *Proc Nat Acad Sci USA*, 2008. 105: p. 11818-11822.

[50] Baba, Y., Y. Osada, and T. Miyashita, The effect of host web complexity on prey-stealing success in a kleptoparasitic spider mediated by locomotor ability. *Anim Behav*, 2012. 83: p. 1261-1268.

[51] Nakata, K., and A. Ushimaru, Differences in web construction behavior at newly occupied web sites between two *Cyclosa* species. *Ethology*, 2004. 110: p. 397-411.

[52] Harmer, A. T. M., T. A. Blackledge, J. S. Madin, and M. E. Herberstein, High-performance spider webs: integrating biomechanics, ecology and behaviour. *J Roy Soc Interf*, 2011. 8: p. 457-471.

[53] Craig, C. L., The ecological and evolutionary interdependence between web architecture and web silk spun by orb web weaving spiders. *Biol J Linn Soc*, 1987. 30: p. 135-162.

[54] Craig, C. L., The significance of spider size to the diversification of spider-web architectures and spider reproductive modes. *Am Nat*, 1987.129: p. 47-68.

[55] Sensenig, A., I. Agnarsson, and T. A. Blackledge, Adult spiders use tougher silk: ontogenetic changes in web architecture and silk biomechanics in the orb-weaver spider. *J Zool*, 2011. 285: p. 28-38.

[56] Anotaux, M., J. Marchal, N. Chaline, L. Desquilbet, R. Leborgne, C. Gilbert, and A. Pasquet, Ageing alters spider orb-web construction. *Anim Behav*, 2012. 84: p. 1113-1121.

[57] Eberhard, W. G., Function and phylogeny of spider webs. *Ann Rev Ecol Syst*, 1990. 21: p. 341-372.

[58] Kuntner, M., S. Kralj-Fiser, and M. Gregoric, Ladder webs in orb-web spiders: ontogenetic and evolutionary patterns in Nephilidae. *Biol J Linn Soc*, 2010. 99: p. 849-866.

[59] Kuntner, M., M. Gregoric, and D. Li, Mass predicts web asymmetry in *Nephila* spiders. *Naturwissenschaften*, 2010. 97: p. 1097–1105.

[60] Vollrath, F., and W. Mohren, Spiral geometry in the garden spider's orb web. *Naturwissenschaften*, 1985. 72: p. 666-667.

[61] Herberstein, M. E., and A. M. Heiling, Asymmetry in spider orb webs: a result of physical constraints? *Anim Behav*, 1999. 58: p. 1241-1246.

[62] Zschokke, S., and K. Nakata, Spider orientation and hub position in orb webs. *Naturwissenschaften*, 2010. 97: p. 43-52.

[63] Rao, D., O. Ceballos Fernandez, E. Castaneda-Barbosa, and F. Diaz-Fleischer, Reverse positional orientation in a neotropical orb-web spider, *Verrucosa arenata*. *Naturwissenschaften*, 2011. 98: p. 699–703.

[64] Graf, B., and W. Nentwig, Ontogenetic change in coloration and web-building behavior in the tropical spider *Eriophora fuliginea* (Araneae, Araneidae). *J Arachnol*, 2001. 29: p. 104-110.

[65] Hesselberg, T., Ontogenetic changes in web design in two orb-web spiders. *Ethology*, 2010. 116: p. 535-545.

[66] Higgins, L. E., Developmental changes in barrier web structure under different levels of predation risk in *Nephila clavipes* (Araneae: Tetragnathidae). *J Ins Behav*, 1992. 5: p. 635-655.

[67] Opell, B. D., Estimating the stickiness of individual adhesive capture threads in spider orb webs. *J Arachnol*, 2002. 30: p. 494-502.

[68] Higgins, L. E., Quantitative shifts in orb-web investment during development in *Nephila clavipes* (Araneae, Nephilidae). *J Arachnol*, 2006. 34: p. 374-386.

[69] Sensenig A., I., Agnarsson and T. A. Blackledge, Behavioural and biomaterial coevolution in spider orb webs. *J Evol Biol*, 2010. 23: p. 1839-1856.

[70] Eberhard, W. G., G. Barrantes, and R. Madrigal-Brenes, Vestiges of an orb-weaving ancestor? The "biogenetic law" and ontogenetic changes in the webs and building behavior of the black widow spider *Latrodectus geometricus* (Araneae Theridiidae). *Ethol Ecol Evol*, 2008. 20: p. 211-244.

[71] Kuntner, M., and I. Agnarsson, Phylogeny accurately predicts behaviour in Indian Ocean *Clitaetra* spiders (Araneae : Nephilidae). *Invert Syst*, 2009. 23: p. 193-204.

[72] Nakata, K., Does ontogenetic change in orb web asymmetry reflect biogenetic law? *Naturwissenschaften*, 2010. 97: p. 1029-1032.

[73] Vollrath, F., B. Madsen, and Z. Shao, The effect of spinning conditions on the mechanics of a spider's dragline silk. *Proc Roy Soc B*, 2001. 268: p. 2339-2346.

[74] Opell, B. D., S. E. Karinshak, and M. A. Siger, Humidity affects the extensibility of an orb-weaving spider's viscous thread droplets. *J Exp Biol*, 2011. 214: p. 2988-2993.

[75] Sahni, V., T. A. Blackledge, and A. Dhinojwala, Changes in the adhesive properties of spider aggregate glue during the evolution of cobwebs. *Sci Rep*, 2011.1: p. 41.

[76] Madsen, B., Z. Shao, and F. Vollrath, Variability in the mechanical properties of spider silks on three levels: interspecific, intraspecific and intraindividual. *Inter J Biol Macromol*, 1999. 24: p. 301-306.

[77] Venner, S., M. C. Bel-Venner, A. Pasquet, and R. Leborgne, Body-mass-dependent cost of web-building behavior in an orb weaving spider, *Zygiella x-notata*. *Naturwissenschaften*, 2003. 90: p. 269-272.

[78] Venner, S., I. Chades, M. C. Bel-Venner, A. Pasquet, F. Charpillet, and R. Leborgne, Dynamic optimization over infinite time horizon: web-building strategy in an orb-weaving spider as a case study. *J Theor Biol*, 2006. 241: p. 725-733.

[79] Blackledge, T. A., and C. M. Eliason, Functionally independent components of prey capture are architecturally constrained in spider orb webs. *Biol Lett*, 2007. 3: p. 456-458.

[80] Crews, S. C., and B. D. Opell, The features of capture threads and orb-webs produced by unfed *Cyclosa turbinata*(Araneae: Araneidae). *J Arachnol*, 2006. 34: p. 427-434.

[81] Townley, M. A., E. K. Tillinghast, and C. D. Neefus, Changes in composition of spider orb web sticky droplets with starvation and web removal and synthesis of sticky droplet compounds. *J Exp Biol*, 2006. 209: p. 1463-1486.

[82] Opell, B. D., G. K. Lipkey, M. L. Hendricks, and S. T. Vito, Daily and seasonal changes in the stickiness of viscous capture threads in *Argiope aurantia* and *Argiope trifasciata* orb webs. *J Exp Zool*, 2009. 311A: p. 217-225.

[83] Blamires, S.J., C. L. Wu, and I. M. Tso, Variations in protein intake induces variations in spider silk expression. *PLoS ONE*, 2012. 7: p. e31626.

[84] Nentwig, W., and H. Rogg, The cross stabilimentum of *Argiope argentata*- nonfunctional or nonspecific stress reaction? *Zool Anz*, 1988. 221: p. 248-266.

[85] Humphreys, W.F., Stabilimenta as parasols: shape construction by *Neogea* sp. (Araneidae) in a successional habitat. *Bull Brit Arach Soc*, 1992. 9: p. 47-52.

[86] Walter, A., C. Westphal, P. Bliss, and R. Moritz, Drinking behaviour of the orb web spider *Argiope bruennichi* (Araneae; Araneidae). *Behaviour*, 2011. 148: p. 1297-1311

[87] Bruce, M. J., A. M. Heiling, and M. E. Herberstein, Spider signals: are web decorations visible to birds and bees? *Biol. Lett*, 2005. 1: p. 299-302.

[88] Cheng, R. C., and I. M. Tso, Signaling by decorating webs: luring prey or detering predators. *Behav Ecol*, 2007. 18: p. 1085-1091.

[89] Blamires, S.J., D. F. Hochuli, and M. B. Thompson, Why cross the web?: affects of decoration spectral properties on the number and type of natural prey captured in an orb spider (*Argiope keyserlingi*) web. *Biol J Linn Soc*, 2008. 94: p. 221-229.

[90] Cheng, R. C., E. C. Yang, E. P. Lin, M. E. Herberstein, and I. M. Tso, Insect form vision as one potential shaping force of spider web decoration design. *J Exp Biol*, 2010. 213: p. 759-768.

[91] Craig, C. L., Limits to learning: effect of predator pattern and colour on perception and avoidance learning by prey. *Anim Behav*, 1994. 47: p. 1087-1099.

[92] Craig, C. L., Predator foraging behavior in response to perception and learning by its prey: interactions between orb-spinning spiders and stingless bees. *Behav Ecol Sociobiol*, 1994. 35: p. 45-52.

[93] Benjamin, S. P., and S. Zschokke, Webs of theridiid spiders: construction, structure and evolution. *Biol J Linn Soc*, 2003. 78: p. 293–305.

[94] Blackledge, T. A., and R. G. Gillespie, Convergent evolution of behavior in an adaptive radiation of Hawaiian web-building spiders. *Proc Nat Acad Sci USA*, 2004. 101: p. 16228–16233.

[95] Harmer, A. T. M., Elongated orb-webs of Australian ladder-web spiders (Araneidae: *Telaprocera*) and the significance of orb-web elongation. *J Ethol*, 2009. 27: p. 453-460.

[96] Gregoric, M., I. Agnarsson, T. A. Blackledge, and M. Kuntner, How did the spider cross the river? behavioral adaptations for river-bridging webs in *Caerostris darwini* (Araneae: Araneidae). *PLoS ONE*, 2011. 6: p. e26847.

[97] Adams, M. R., Choosing hunting sites: web site preferences of the orb weaver spider, *Neoscona crucifera*, relative to light cues. *J Ins Behav*, 2000. 13: p. 299-305.

[98] Eberhard, W. G., Effects of gravity on temporary spiral construction by *Leucauge mariana* (Araneae: Araneidae). *J Ethol*, 1987. 5: p. 29-36.

[99] Eberhard, W. G., and T. Hesselberg, Cues that spiders (Araneae: Araneidae, Tetragnathidae) use to build orbs: lapses in attention to one set of cues because of dissonance with others? *Ethology*, 2012. 118: p. 610-620.

[100] Landolfa , M. A., and F. G. Barth, Vibrations in the orb web of the spider *Nephila clavipes*: cues for discrimination and orientation. *J Comp Physiol A*, 1996. 179: p. 493-508.

[101] Krink, T., and F. Vollrath, A virtual robot to model the use of regenerated legs in a web-building spider. *Anim Behav*, 1999. 57: p. 223-232.

[102] Sandoval, C. P., Plasticity in web design in the spider *Parawixia bistriata*: a response to variable prey type. *Funct Ecol*, 1994. 8: p. 701–707.

[103] Nakata, K., Prey detection without successful capture affects spider's orb-web building. *Naturwissenschaften*, 2007. 94: 853-857.

[104] Nakata, K., Spiders use airborne cues to respond to flying insect preators by building orb-web with fewer silk thread and larger silk decorations. *Ethology*, 2008. 114: p. 686-692.

[105] Hughes, D., Pathways to understanding the extended phenotype of parasites in their hosts. *J Exp Biol*, 2013. 216: p. 142-147.

[106] Thomas, F., A. Schmidt-Rhaesa, G. Martin, C. Manu, P. Durand, and F. Renaud, Do hairworms (Nematomorpha) manipulate the water seeking behaviour of their terrestrial hosts? *J Evol Biol*, 2002. 15: p. 356–361.

[107] Andersen, S. B., S. Gerritsma, K. M. Yusah, D. Mayntz, N. L. Hywel-Jones, J. Billen, J. .J. Boosma, and D. P. Hughes, The life of a dead ant: the expression of an adaptive extended phenotype. *Am Nat*, 2009. 174: p. 424-433.

[108] Eberhard, W. G., Spider manipulation by a wasp larva. A parasitic wasp forces its host to weave a special web for its own ends. *Nature*, 2000. 406: p. 255-256.

[109] Korenko, S., and S. Pekar, A parasitoid wasp induces overwintering behaviour in its spider host. *PLoS ONE*, 2011. 6: p. e24628.

[110] Fincke, O. M., L. Higgins, and E. Rojas, Parasitism of *Nephila clavipes* (Araneae, Tetragnathidae) by an ichneumonid (Hymenoptera, Polysphinctini) in Panama. *J Arachnol*, 1990. 18: p. 321–329.

[111] Eberhard, W. G., Under the influence: webs and building behaviour of *Plesiometa argyra* (Araneae, Tetragnathidae) when parasitized by *Hymenoepimecis argyraphaga* (Hymenopera, Ichneumonidae). *J Arachnol*, 2001. 29: p. 354–366.

[112] Gonzaga, M. O., and J. F. Sobczak, Parasitoid-induced mortality of *Araneus omnicolor* (Araneae, Araneidae) by *Hymenoepimecis* sp. (Hymenoptera, Ichneumonidae) in southeastern Brazil. *Naturwissenschaften*, 2006. 94: p. 223–227.

[113] Sobczak, J. F., A. P. S. Loffredo, A. M. Penteado-Dias, and M. O. Gonzaga, Two new species of *Hymenoepimecis* (Hymenoptera: Ichneumonidae: Pimplinae) with notes on their spider hosts and behaviour manipulation. *J Nat Hist*, 2009. 43: p. 2691–2699.

[114] Eberhard, W. G., New types of behavioral manipulation of host spiders by a parasitoid wasp. *Psyche*, 2010: p. 950614.

[115] Gonzaga, M. O., J. F. Sobczak, A. M. Penteado-Dias, and W. G. Eberhard, Modification of *Nephila clavipes* (Araneae Nephilidae) webs induced by the parasitoids *Hymenoepimecis bicolor* and *H. robertsae* (Hymenoptera Ichneumonidae). *Ethol Ecol Evol*, 2010. 22: p. 151–165.

[116] Tanaka, S., and N. Ohsaki, Behavioral manipulation of host caterpillars by the primary parasitoid wasp *Cotesia glomerata* (L.) to construct defensive webs against hyperparasitism. *Ecol Res*, 2006. 21: p. 570–577.

[117] Grossman, A. H., A. Janssen,, E. F. de Brito, E. G. Cordeirgo, F. Colares, J. Oliveira Fonseca, E. R. Lima, E. R. Pallini, and M. W. Sabelis, Parasitoid increases survival of its host pupae by inducing hosts to fight predators. *PLoS ONE*, 2008. 3: p. e2276.

[118] Boutry, C., and T. A. Blackledge, Biomechanical variation of silk links spinning plasticity to spider web function. *Zoology*, 2009. 112: p. 451-460.

[119] Blackledge, T. A. Spider silk: a brief review and prospectus on research linking biomechanics and ecology in draglines and orb webs. *J Arachnol*, 2012. 40: p. 1-12.

[121] Laland, K. N., and K. Sterelny, Seven reasons (not) to neglect niche construction. *Evolution*, 2006. 60: p. 1751-1762.

[122] Laland, K. N., J. Odling-Smee, and M. W. Feldman, Evolutionary consequences of niche construction and their implications for ecology. *Proc Nat Acad Sci USA*, 1999. 96: p. 10242-10247.

[123] Day, R. L., K. N. Laland, and J. Odling-Smee, Rethinking adaptation: the niche construction perspective. *Persp Biol Med*, 2003. 46: p. 80-95.

[124] Venner, S., and J. Casas, Spider webs designed for rare but life-saving catches. *Proc Roy Soc B*, 2005. 272: p. 1587-1592.

[125] Thompson, J. N., The coevolving web of life. *Am Nat*, 2009. 173: p. 125-140.

[126] Whitehouse, M. E. A., and Y. Lubin, The functions of societies and the evolution of group living: spider societies as a test case. *Biol Rev*, 2005.80: p. 347-361.

[127] Kearney, M., and W. Porter, Mechanistic niche modelling: combining physiological and spatial data to predict species ranges. *Ecol Lett*, 2009. 12: p. 334-350.

[128] Berg, M. J., and J. Ellers, Trait plasticity in species interactions: a driving force of community dynamics. *Evol Ecol*, 2010. 24: p. 617-629.

[129] Walther, G. R., E. Post, P. Convey, A. Manzel, C. Parmesan, T. J. C. Beebee, J. M. Fromentin, O. Hoegh-Guldberg, and F. Bairlein, Ecological responses to recent climate change. *Nature*, 2002. 416: p. 389-395.

[130] Dondale C. D., H .J. Redner , and L. LeSage, A comb-footed spider *Achaearanea tabulata*, new to the fauna of Canada (Araneae: Theridiidae). *J Arachnol*, 1994. 22: 176–178.

[131] Levi H. W., Two new spiders of the genera *Theridon* and *Achaearanea* from North America (Araneae: Theridiidae). *Trans Am Micros Soc*, 1980. 99: p. 334-337.

[132] Roberts, M. I., *The spiders of Great Britain and Ireland. volume 1*. 1985, London: Harper Collins.

[133] Yoshida, H. A new species of the genus *Achaearanea* (Araneae: Theridiidae) from Japan. *Acta arachnol*, 1983. 32: 37-42.

[134] Nentwig, W., The prey of spiders. *Ecophysiology of spiders*. ed. W. Nentwig. 1987, Berlin: Springer-Verlag. p. 249-263.

[135] Hansell, M. H., Nest building as a facilitating and limiting factor in the evolution of eusociality in the Hymenoptera. *Oxford surveys in evolutionary biology volume 4*. eds. P. H. Harvey, and L. Partridge. 1987, Oxford: Oxford University Press. p. 155-181.

[136] Blackledge, T. A., N. Scharff, J. A. Coddington, T. Szuts, J. W. Wenzel, C. Y. Hayashi, and I. Agnarsson, Reconstructing web evolution and spider diversification in the molecular era. *Proc Nat Acad Sci USA*, 2009. 106: p. 5229-5234.

[137] Blamires, S. J., C. L.Wu, T. A. Blackledge, and Tso, I. M. Environmentally-induced post-spin property changes in spider silks: influence of web type, spidroin composition and ecology. *Biol J Linn Soc*, 2012.106: p. 580-588.

[138] Stuart, A. E., and F. F. Hunter, End-products of behaviour versus behavioural characters: a phylogenetic investigation of pupal cocoon

construction and form in some North American black flies (Diptera: Simuliidae). *Ecol Entomol*, 1998. 23: p. 387-398.

[139] Elices, M., G. R. Plaza, M. A. Arnedo, J. Perez-Rigueiro, F. G. Torres, and G. V. Guinea, Mechanical behavior of silk during the evolution of orb-web spinning spiders. *Biomacromolecules*, 2009. 10: p. 1904-1910.

[140] Blackledge, T. A., J. Perez-Riguero, G. R.Plaza, B. Perea, A. Navarro, G. V. Guinea, and M. Elices, Sequential origin in the high performance properties of orb spider dragline silk. *Sci Rep*, 2012. 2: p. 782.

[141] Blamires, S.J., C. L.Wu, T. A. Blackledge, and I. M. Tso, Post-secretion processing influences spider silk performance. *J Roy Soc Interf*, 2012. 9: p. 2479-2487.

[142] McGeoch, M. A., The selection, testing and application of terrestrial insects as bioindicators *Biol Rev*, 1998. 73: p. 181-201.

[143] Marc, P., A. Canard, and F. Ysnel, Spiders (Araneae) useful for pest limitation and bioindication. *Ag Ecosyst Environ*, 1999. 74: p. 229-274.

[144] Wilczek, G., Apoptosis and biochemical biomarkers of stress in spiders from industrially polluted areas exposed to high temperature and dimethoate. *Comp Biochem Physiol C*, 2005. 141: p. 194-206.

[145] Nummelin, M., M. Lodenius, E. Tulisalo, H. Hirvonen, and T. Alanko, Predatory insects as bioindicators of heavy metal pollution. *Environ Poll*, 2007. 145: 339-347.

[146] Eraly, D., F. Hendrickx, T. Backeljau, L. Bertvoets, and L. Lens, Direct and indirect effects of metal stress on physiology and life history variation in field populations of a lycosid spider. *Ecotoxicol Environ Safety*, 2011. 74: p. 1489-1497.

[147] Driessen, M. M., Enhancing conservation of the Tasmanian glow-worm, *Arachnocampa tasmaniensis* Ferguson (Diptera: Keroplatidae) by monitoring seasonal changes in light displays and life stages. *J Ins Cons*, 2010. 14: p. 65-75.

[148] Magoshi, J., T. Tanaka, H. Sasaki, M. Kobayashi, Y. Magoshi, H. Tsuda, M. A. Becker, S. I. Inoue, and K. Ishimaru, Uptake of atmospheric carbon dioxide into silk fiber by silkworms. *Biomacromolecules*, 2003. 4: p. 778-782.

[149] Zax, D. B., D. E. Armanios, S. Horak, C. Malowniak, Z. Yang, Varaition of mechanical properties with amino acid content in the silk of *Nephila clavipes*. *Biomacromolecules*, 2004. 5: p. 732-738.

[150] Zhou, L., X. Chen, Z. Shao, Y. Huang, and D. P. Knight, Effect of metallic ions on silk formation in the mulberry silkworm, *Bombyx mori*. *J Phys Chem B*, 2005. 109: p. 16937-16945.

[151] Hose, G. C., J. M. James, and M. R. Gray, Spider webs as environmental indicators. *Environ Poll*, 2002. 120: p. 725-733.

In: Spiders
Editor: Marq Santerre

ISBN: 978-1-62618-517-3

Chapter 3

LIPOPROTEINS OF SPIDERS: STRUCTURE AND FUNCTION

C. Fernando Garcia**, Aldana Laino and Monica Cunningham

INIBIOLP (CONICET-UNLP)- Fac. de Cs. Naturales y Museo, UNLP. La Plata, Argentina

ABSTRACT

It is well known that the Class Arachnida is one of the arthropod groups with major diversity in the world. There are around 100 families belonging to the Order Araneae; over 42,000 species from different geographical areas have been described. In the last few years biochemical/physiological studies in some species of spiders: *Polybetes pythagoricus, Latrodectus mirabilis, Eurypelma californicum, Cupiennius salei, Nephila inaurata* among others have significantly increased. Many of them are based on the venom characterization, but only a few of them are related to basic molecules such as lipids, proteins or carbohydrates.

Lipids are transported from the tissue of origin to those tissues where they are either stored or metabolized. In an aqueous environment as the hemolymph they must be associated to proteins, forming lipoproteins which may be more complex generating the so-called glycolipoproteins.

* Corresponding author E-mail: cfgarcia1123@yahoo.com.ar.

Within the Class Arachnida, different lipoproteins were identified and characterized in the Orders Araneae, Acarina, Scorpionida and Solpugida. In only three species of spiders, *Polybetes pythagoricus* and *Latrodectus mirabilis* (Suborder Araneomorphae) and *Eurypelma californicum* (Suborder Mygalomorphae), lipoproteins of different density were purified and studied. *P. pythagoricus* belongs to the Sparassidae family, and it is only found in South America, while *L. mirabilis* belongs to the Theridiidae family and is distributed in the center and south of Argentina and Uruguay. *E. californicum* belongs to the Theraphosidae family that is present in the American continent, mainly in North America. Lipoproteins of high density ranging from 1.07 to 1.14 g/ml have been detected as the major circulating lipoproteins in these species. Concerning the lipid composition of Araneomorphae, the major lipids are phosphatidylcholine, free fatty acids and triacylglycerides (being this one the major energetic lipid). As to Myghalomorphae *E. californicum*, this lipoprotein of high density named lipophorin as in insects, mainly contains phospholipids and diacylglycerol (the major energetic lipid). Another high density lipoprotein named HDL2 is found in *P. pythagoricus* and *L. mirabilis*, while *P. pythagoricus* and *E. californicum* have a third lipoprotein of very high density (VHDL) in hemolymph. It is important to remark that the hemocyanin in *P. pythagoricus* and *L. mirabilis*, in addition to its classical role as respiratory pigment, plays functions as apolipoprotein. Lately, *P. pythagoricus* has been used as a model in order to further study hemolymph lipoproteins (HDL1, HDL2 and VHDL). For the first time, egg glycolipoproteins of high and very high density (LV1 and LV2) which are responsible for the embryo nutrition, were identified in class Arachnida. Phospholipids are the major structural lipids. Oleic acid is the major component in both lipoproteins. These particles are spheroidal and their diameter ranges between 20 nm in LV1 and 27 nm in LV2.

Recently, we have reported data on the secondary and tertiary structure of VHDL by using spectroscopy and sequencing methods.

Despite the fact that biochemical and physiological studies dealing with lipoproteins have increased, further studies are necessary to elucidate the dynamics of proteins and lipids in arachnids.

It is well known that the Class Arachnida is one of the arthropod groups with major diversity in the world. There are around 100 families belonging to the Order Araneae; over 42,000 species from different geographical areas have been described. Besides this group has reached major adaptive diversity in arachnids; they are distributed in all the habitats where life is possible, except in the air and sea at large from the sea level up to 7,000 m in Mount Everest. A few species were able to be adapted to semiaquatic life, but only one could survive under water despite the aerial respiration. There are spiders of different size, small body of 2 to 10 mm long to the so called tarantula which body may be 80-90 mm long. Most spiders live away from man, but they may go into their houses to get insects for feeding. A few species of spiders are poisonous for the man, and they may even be deadly.

However, spiders are important from the ecological point of view as they play a key role in the trophic chain. On the one hand they are food for some animals, and on the other hand they exert a natural control over some populations such as insects. This fact is essential to maintain the biologic equilibrium in biocenosis and ecosystems (Hoffmann, 2003).

Currently, man is more interested in these arthropods. In the last few years structural, genetic and biochemical aspects of spiders have been investigated, mainly concerning silk, poison and environment.

Thus, novel studies on these organisms showed they are beneficial for man, and their silk could be utilized as biomaterial (Schacht and Scheibel, 2011); it could be applied in the field of biomedicine either surgery material (Aparicio et al., 2010), or copying bioadhesives (Sahni et al., 2010). Silk recombinant proteins have been studied as possible platforms for administering non viral genes to tumor cells (Numata et al., 2011).

Also spiders play an important role in agriculture by controlling crop cultures (Cheli et al., 2011; Armendano and González, 2010; González et al., 2009). There are many reports dealing with different classes of pharmacologically active compounds which are present in arachnid venoms (Beraud and Chandy, 2011).

Biochemical and physiological studies reported essential information to get an insight into different features of these organisms such as biology, reproduction, nutrition, molecular-environmental connections, etc. Lipids and proteins are important components due to their structural and energy role as well as their composition that may change according to sex, age and nutritional state and reproductive.

Lipids are transported in aquatic media by binding with proteins called apolipoproteins. Sometimes, more complex structures as glycolipoproteins with an additional component, carbohydrates, may be generated.

In general, arthropod lipoproteins are classified in accordance to that pattern developed by Scanu et al., (1975). This classification is based on hydration density of lipoprotein particles; only lipoproteins of high density (HDL) and very high density (VHDL) were found in arachnids.

Concerning their function lipoproteins are classified as non related and related to sex.

The first ones called hemolymph lipoproteins are responsible for transporting lipids from their site of origin to the different organs.

In some arachnids hemocyanin was shown to play a "secondary" function as lipid carrier in addition to that of oxygen carrier, leading to a subdivision of hemolymph lipoproteins into: without hemocyanin and with hemocyanin as apolipoproteins (see below).

The second ones are reproduction-related lipoproteins; they are only present in females under reproduction or vitellogenic stage (vitellogenins) and those found in ovary and/or eggs are named lipovitellins.

Figure 1. An adult female specimen of *Polybetes pythagoricus* (Araneae, Sparassidae).

HEMOLYMPH LIPOPROTEINS WITHOUT HEMOCYANIN AS APOLIPOPROTEIN

Arthropod lipoproteins have been isolated and described in crustacean and insects. Several studies were done on crustaceans, focusing mainly the aquatic ones. In 2000 Yepiz-Placencia and collaborators published a reference review on peneid lipoproteins where their study is mainly based on protein part of lipoproteins (Yepiz-Plascencia et al., 2000).

Concerning insects, lipophorin is the principal lipoprotein in this group of arthropods. It has two apolipoproteins of 250 kDa and 80 kDa molecular weight, but in some insects a third apolipoprotein may be found (Shapiro et al., 1988; Bianchi et al., 1987; Ryan et al., 1984; Kawooya et al., 1984).

In general, reports dealing with spiders are poor. Haunerland and Bowers (1987) were the first to depict tarantula *Eurypelma californicum* (Ausserer, 1871) which is, now called *Aphonopelma hentzi* (Girard, 1852) in taxonomy publications (Wolfgang, 2012) (Araneae, Theraphosidae). They isolated and identified two lipoproteins without hemocyanin by ultracentrifugation and specific dyeing for lipids; they were named HDL1 and VHDL. The first one has a molecular weight of 500 kDa and two subunits of 250 and 80 kDa whereas the second one of 220 kDa has two subunits of 120 and 105 kDa (Haunerland and Bowers, 1987).

Later, Stratakis et al., (1993) reported further information about VHDL with a native molecular weight of 380/420 kDa, and two subunits of 110 and 95 KDa (Stratakis et al., 1993). In 1994 Cunningham et al. described lipoproteins in the spider *Polybetes pythagoricus* (Holmberg, 1875) (Araneae, Sparassidae) (Figure 1) where one HDL lipoprotein named HDL1 with two subunits of 250 and 80 kDa were found (Cunningham et al., 1994). Then, these authors isolated and characterized hemolymphatic lipoproteins of the spider *Latrodectus mirabilis* (Holmberg, 1876) (Araneae, Theridiidae) (Figure 2).

The lipoprotein without hemocyanin was named HDL1 which is composed of two subunits of 103 and 90 kDa (Cunningham et al., 2000). As stated before, HDL apolipoproteins of *E. californicum* as well as those of HDL1 in *P. pythagoricus* have the same molecular weight as that of lipophorin in insects (Table 1). It is well known that spiders are phylogenetically quite distant from insects, though apolipoproteins of lipophorins have the same molecular weight. On the other hand, the similarity between acarian and crustacean lipoproteins is remarkable; there is, however, no plausible taxonomic explanation for this fact (Haunerland and Bowers, 1987; Steward and Fisher, 1985).

In general, HDLs have 20% to 40% lipids and VHDL contain 8 to 16% (Table 1). Lipoproteins contain phospholipids as major structural lipids (32% to 60%), coinciding with data dealing with lipoproteins of invertebrates. Concerning energy lipids of HDL in *E. californicum,* a high content of diacylglycerols (DAG) (39%) was found; a similar fact that was observed in lipophorins (Haunerland and Bowers, 1989).

In *P. pythagoricus* and *L. mirabilis*, triacylglycerides (TAG) represent the major energy lipids (16% and 24%, respectively) (Cunningham et al., 2000; Cunningham et al., 1994). These lipoproteins contain another predominant lipid, the free fatty acid (FFA), ranging from 34% in VHDL of *E. californicum*, and 33% in HDL1 of *L. mirabilis* to 17% in HDL1 of *P. pythagoricus* (Cunningham et al., 2000; Cunningham et al., 1994; Haunerland and Bowers, 1987).

Figure 2. An adult specimen of *Latrodectus mirabilis* (Araneae, Theridiidae) (Photograph kindly donated by Francisco Giambelluca).

Table 1. Hemolymph lipoproteins without hemocyanin as apolipoprotein

Species Lipoproteins	Density gr/ml	% lipids	% proteins	Amount of lipids transported %	Lipid clases %	Particle mass kDa	Apolipoproteins MW kDa
E. californicum HDL (1)	1.10/1.14	40/44	60/56	N/D	DAG 39 PL 32 TAG 11	500	250 80
E. californicum VHDL (1)	1.28	8	92	N/D	PC, TAG, FFA (n/d)	220	120 105
E. californicum VHDL (2)	1.23	16	84	N/D	PL 45 FFA 34 TAG 7	380/420	110 95
L. mirabilis HDL 1 (3)	1.13	20	80	80	PL 34 FFA 33 TAG 24	N/D	103 90
P. pythagoricus HDL 1 (4)	1.13	28	72	26	PL 60 AGL 17 TAG 16	N/D	250 80

(1) Haunerland and Bowers (1987); (2) Stratakis et al., (1993); (3) Cunningham et al., (2000); (4) Cunningham et al., (1994); (5) Laino et al., (2011).

DAG; diacylglycerol; PL; phospholipids; TAG; triacylglycerol; PC; phosphatidylcholine; FFA; free fatty acids; COL; cholesterol; HC; hydrocarbons; PE; phosphatidyl ethanolamine. EE; sterols esters.

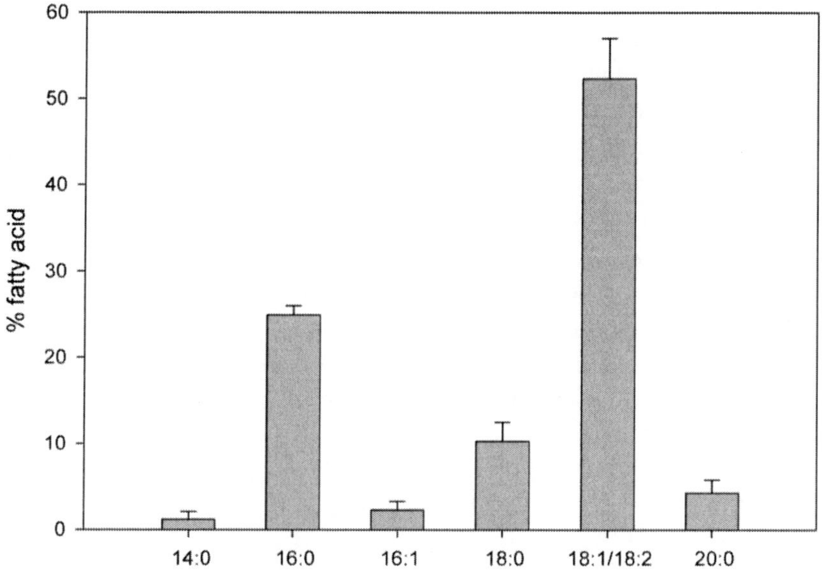

Figure 3. Fatty acid composition of VHDL from *E. californicum* (Stratakis et al., 1993).

Stratakis et al., (1993) have reported new data on the VHDL structure of the tarantula *E. californicum,* indicating that its major fatty acids were 18:1 and 18:2 (52%) (Figure 3). Similar results were reported by Schartau and Leidescher (1983) when investigating whole hemolymph.

HEMOLYMPH LIPOPROTEINS WITH HEMOCYANIN AS APOLIPOPROTEIN

In nature there are different respiratory pigments, hemocyanin is the only one present in spiders. It is a copper-containing protein which is characterized by its bluish colour. In arthropods, it may be found in crustaceans and chelicerates (Cunningham et al., 2007; Decker and Jaenicke, 2004; Jaenicke and Decker, 2004; Bridges, 2001; Cunningham et al., 2000; Cunningham et al., 1999; Decker and Rimke, 1998; Terwilliger, 1998; Cunningham and Pollero, 1996). Its main function is oxygen transport, but other functions have been described such as phenoloxidase and antimicrobial activity as well as lipid transport as apolipoprotein. This novel function has been depicted also in two species of molluscs. Arthropod hemocyanins are multisubunit proteins with a molecular mass of about 70-75 kDa per polypeptide chain. Subunits are arranged as hexamers or multiples of hexamers to which the native aggregation level is species-specific (Voit et al., 2000; Markl, 1986; Van Bruggen, 1980).

The interaction of hemocyanin and lipids has been described for the first time by Zatta (1981), who reported the presence of small amounts of phospholipids that were transported by hemocyanin in the crab *Carcinus maeñas.* He also stated that these lipids played a stabilizing role within the apolipoprotein structure. The first report on the probable action of hemocyanin in lipid transport was published by Heras and Pollero (1992; 1990) in the mollusc *Octopus tehuelchus*. In 1995, Hall et al. (Hall et al., 1995) showed that hemocyanin of the freshwater crab *Pacifastacus leniusculus* was associated to a considerable part of total lipids in hemolymph.

Hemocyanin-containing lipoproteins have been isolated, identified and characterized in only two species of spiders *L. mirabilis* and *P. pythagoricus*. In *L. mirabilis* one lipoprotein of high density 1.19-1.20 gr/ml named HDL2 was determined; it only contains1% lipids such as FFA, phosphatidylcholine (PC) and cholesterol (COL) (28, 25 and 20%, respectively) (Table 2). Due to its abundance in hemolymph this lipoprotein was found to transport 19% hemolymphatic total lipids. The hemocyanin monomer was present in different protein complexes that were analyzed by chromatography and electrophoresis (Cunningham and Pollero, 1996) (Table 2).

In *P. pythagoricus* two hemocyanin-containing lipoproteins were described; one lipoprotein of high density called HDL2 of 1.18-1.20 gr/ml density has 3.5 % lipids and its abundance in hemolymph reveals that it transports 25% of hemolymph total lipids. Its lipid composition contains PC, FFA and other phospholipids (22, 18, 17% respectively).

The other lipoprotein called VHDL evidenced 1.21-1.24 g/ml density, a similar lipid and protein composition to that of HDL2 (Table 2), and it transports more lipids in hemolymph (49%). Recently, we have demonstrated for the first time, that the midgut diverticula of arachnids is a primary storage site and a major lipid metabolic center involved in the uptake and mobilization of lipids.

Also it clearly demonstrated that HDL1 and innovatively the VHDL from *P. pythagoricus* participate in the lipid transport from and towards tissues (Laino et al., 2009). Because of this new and surprising function of VHDL from *P. pythagoricus*, we have got an insight into the structure of VHDL.

Table 2. Hemolymph lipoproteins with hemocyanin as apolipoprotein

Species Lipoproteins	Density gr/ml	% lipids	% proteins	Amount of lipids transported %	Lipid clases %	Particle mass kDa	Apolipo-proteins MW kDa
L. mirabilis HDL 2 (3)	1.19/1.20	1	99	19	FFA 28 PC 25 COL 20 HC 14 TAG 8	440 121 90 70	76 67
P. pythagoricus VHDL (4)	1.21/1.24	3	97	49	PC 40 FFA 14 PE 14 TAG 13	500 420 70	121 105 70
P. pythagoricus HDL 2 (4)	1.28/1.20	3.5	96.5	25	PC 22 FFA 18 Others PL 17 TAG 15	500 420 70	121 105 70

Reference in table 1.

Structural characteristics of VHDL from *P. pythagoricus* were studied by CD spectroscopy. Available information on the secondary structure of spiders and related groups is very scarce. In the *P. pythagoricus* lipoprotein we observed a slightly higher content of β-sheet than that of α-helix, 29% and 21.6% respectively; and 22.7% turns (Figure 4); which is similar to that observed by Dolashka-Angelova et al., (2005) in the crustacean *Carcinus aestuarii*, in which the hemocyanin yields the following averaged values of secondary structure components: α-helix 16.0%; β-sheet 30.1%; turns 20.0%; random coil 34.0%. In other spider, only the hemocyanin from *E. californicum* has been characterized. In contrast with that of *P. pythagoricus* it revealed a content of α-helix 25.2%, 21.6% β-sheet, 22.4% turns and 28.5 unordered. In the hemocyanin of scorpion *Pandinus imperator* the percentage of α-helix on β-sheet is slightly higher again. The scorpion *Pandinus imperator* has 26.5% helix, 20.6% sheet, 22.6% turns and 28.6 unordered (Baird et al., 2007).

This lipoprotein from *P. pythagoricus* was subjected to SDS-PAGGE under oxidizing and reducing conditions (Figure 5). This study showed no differences between the electrophoresis profiles obtained either in the presence or absence of β-mercaptoethanol, indicating the absence of intermolecular disulfide bridges in the lipoprotein particle (Figure 5). Molecular weights of both the monomer and those of non-respiratory proteins were equal in both conditions, 70, 105 and 121 kDa, respectively.

The highest susceptibility of the 105 and 121kDa proteins to trypsinolysis suggests that these subunits are more exposed to the aqueous medium (Figure 6). These two subunits are part of the hexameric structure (Cunningham and Pollero, 1996). Perhaps the trypsinolysis pattern indicates that the lipid transport is involved in these subunits. It is important to point out the only difference existing between VHDL and HDL2 of *P. pythagoricus* is related to the small amount of non respiratory proteins 121 y 105 kDa present in HDL2, and this one could be responsible for the fact that it is not involved in the dynamics of lipids (Laino et al., 2009).

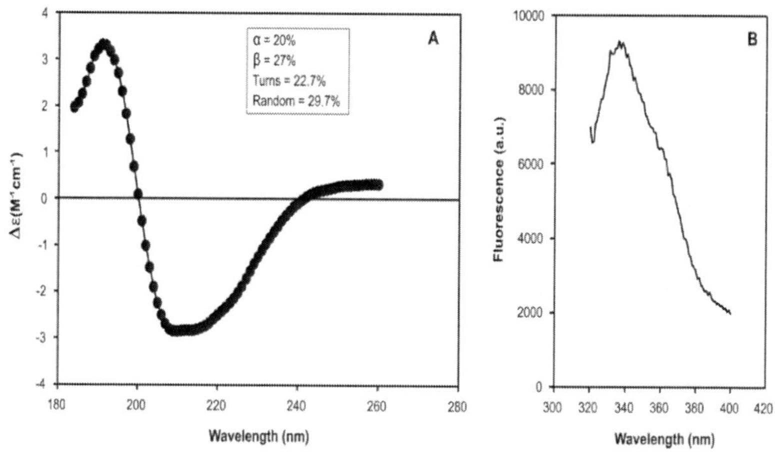

Figure 4. (A) Far-UV circular dichroism spectrum of VHDL from *P. pythagoricus*. (B) Intrinsic fluorescence emission spectrum of VHDL.

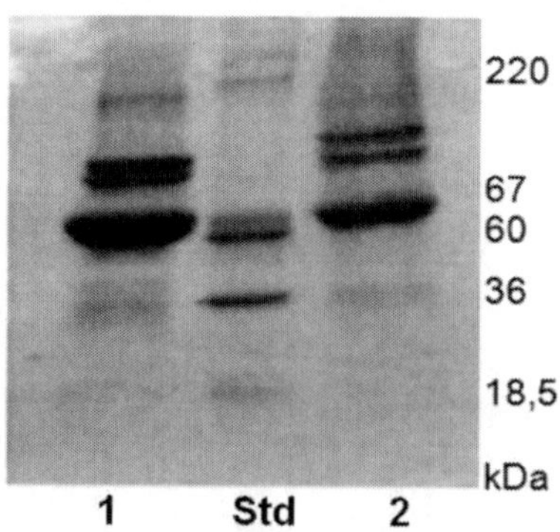

Figure 5. VHDL of *P. pythagoricus*. SDS gel electrophoresis (4-23% acrylamide gradient slab). Lanes 1 and 2. VHDL without and with β-mercaptoethanol.

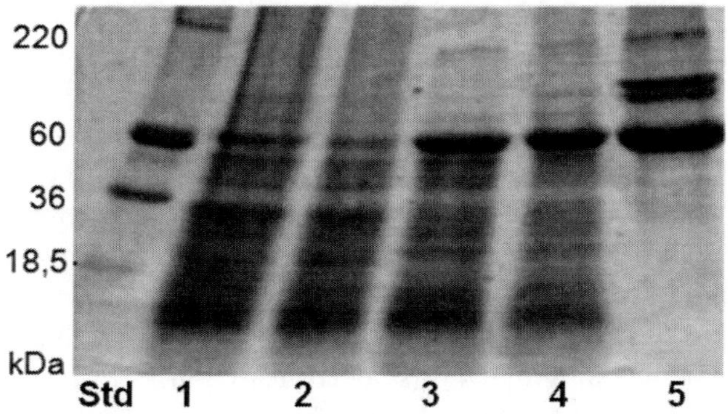

Figure 6. VHDL of *P. pythagoricus.* SDS gel electrophoresis. Lanes 1 and 2: VHDL with trypsin (1:20) for 30 and 60 min of incubation respectively. Lanes 3 and 4: VHDL with trypsin (1:50) for 30 and 60 min of incubation respectively. Lane 5: VHDL control.

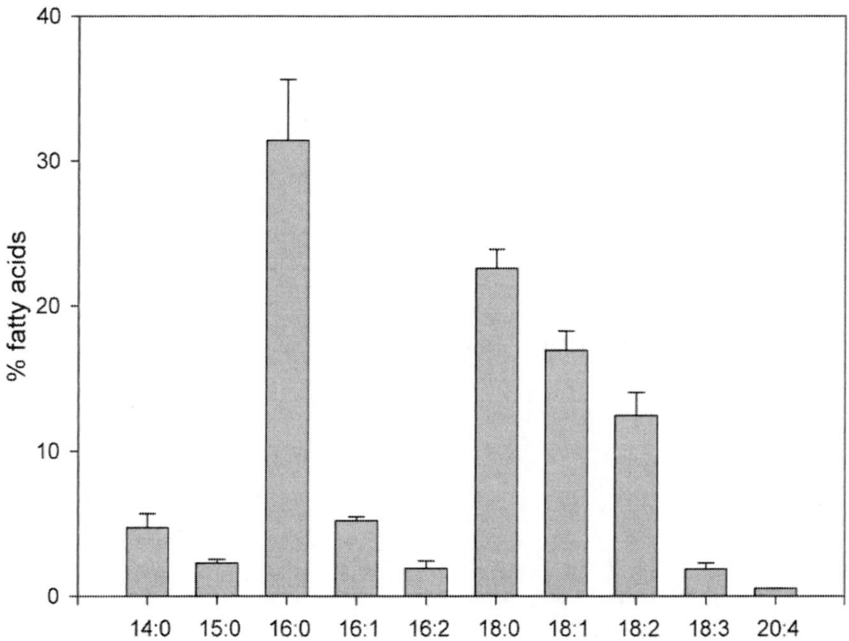

Figure 7. Fatty acid composition of subfraction II from VHDL of *P. pythagoricus.* Data on fatty acid classes are expressed as weight percentage and quantified by gas-liquid chromatography (GLC). Values are the mean +/- 1 SD of three independent analyses.

Fatty acid composition was analyzed by gas chromatography in the subfraction where lipids of the VHDL lipoprotein of *P. pythagoricus* are mainly transported. Upon these studies it was proved that 16:0 and 18:0 (31 and 22%) were the predominant saturated fatty acids while 14:0 and 15:0 were found to be in a minor proportion (less than 6%). The major insaturated acids were 18:1 and 18:2 (17% and 12%, respectively); polyunsaturated acids represented less than 6% (Figure 7) (Laino, Garcia and Cunningham, data not published). It is important to note that this is the first report on the fatty acid composition in lipoproteins with hemocyanin of the Order Araneae which are present in most circulating lipids of spiders, mainly in *P. pythagoricus* due to the fact that 96% of lipids transported are hydrolyzable (Cunningham and Pollero, 1996). Taking into account that the intestinal diverticula is an organ analog to the fat body in insects and the hepatopancreas in crustaceans, it may be involved in the lipid biosynthesis of spiders. For this reason fatty acids in the intestinal diverticula of *P. pythagoricus* (Laino et al., 2009) were characterized (Figure 8). It can be noted that the same fatty acid composition is found in VHDL, which is dominated by 18 and 16 carbon fatty acids, 18:1, 18:2, 16:0. The spectrum corresponds, in general to a "terrestrial pattern" where fatty acids of the n-6 family prevail over those of the n-3 family.

It must be noted that these studies were performed on complete lipoprotein particles, i.e. considering the hemocyanin as apolipoprotein. Generally, previous studies on arachnid hemocyanins have treated these proteins in a free form, disregarding the associated lipid component.

EGG LIPOPROTEINS. LIPOVITELLINS

In oviparous species most nutrients are found in the vitellus, part that is indispensable for the embryo development. Vitellus in invertebrates is composed of lipids, proteins and hydrocarbons, but associated pigments may be also found. Most eggs store nutrients in the yolk in the form of lipoprotein complex called lipovitellin or as lipid droplets scattered throughout the cytoplasm. Vitellogenesis is the process of vitellus accumulation and in most invertebrates it involves massive synthesis of vitellogenin (these lipoproteins are only present in females throughout the reproductive cycle) and its deposition in the oocyte as vitellins or lipovitellins.

Lipovitellins aroused interest in the field of investigation since they can determine embryo and species survival. Poor information on lipoproteins in arachnids makes it difficult to understand structure-function correlation.

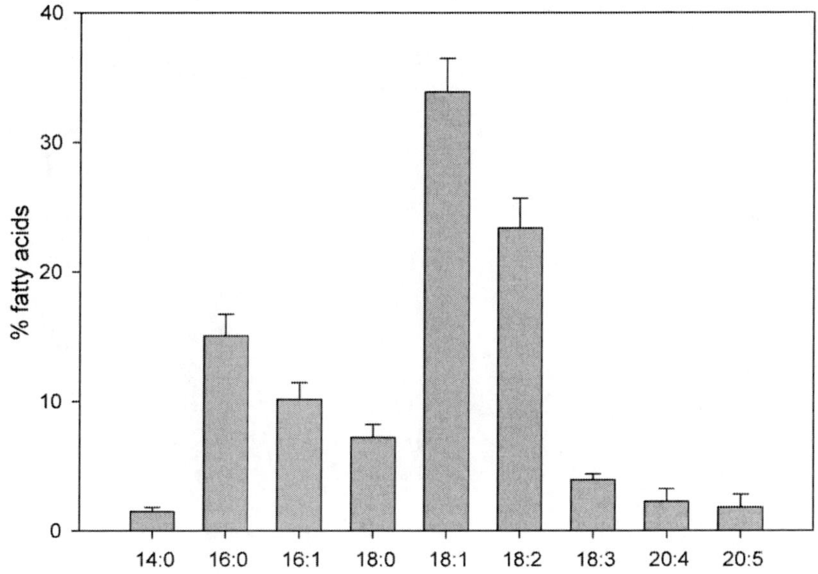

Figure 8. Fatty acid composition of midgut diverticula of *P. pythagoricus*. Data on fatty acid classes are expressed as weight percentage and quantified by GLC.

Although arthropod terrestrial vitellins are well characterized in insects (Tufail and Takeda, 2008; Salerno et al., 2002; Chino, 1997; Dhadialla and Raikhel, 1990) as well as in aquatic arthropod as crustaceans (Walker et al., 2006; García et al., 2006; Chen et al., 2004; Kawazoe et al., 2000; Lubzens et al., 1997), data concerning arachnids are poor, with the exception of acari which have been reported in several works. In this Order, lipovitellins have been characterized in several species which have either vitellins or vitellogenins that are complex hemo-glyco-lipo-phospho-proteins, a structure mainly assumed to be related with their haematophagous habits. Their lipid and glucidic composition is quite similar to that of vitellins in insects (Boldbaatar et al., 2010; James and Oliver, Jr., 1997; Rosell and Coons, 1992; Schriefer, 1991; Shanbaky et al., 1990; Dhadialla, 1986; Chinzei et al., 1983; Boctor and Kamel, 1976; Tatchell, 1971).

The importance given to the studies on lipovitellins has been increasing in the last years because of their relationship with other lipoproteins and their key role in the reproductive strategies of invertebrates. In this regard, Baker determined that the vitellogenin of some invertebrates (Nematoda and Insecta) contains domains homologs to those of apoB-100 of the LDL and to human lipase (Baker, 1988).

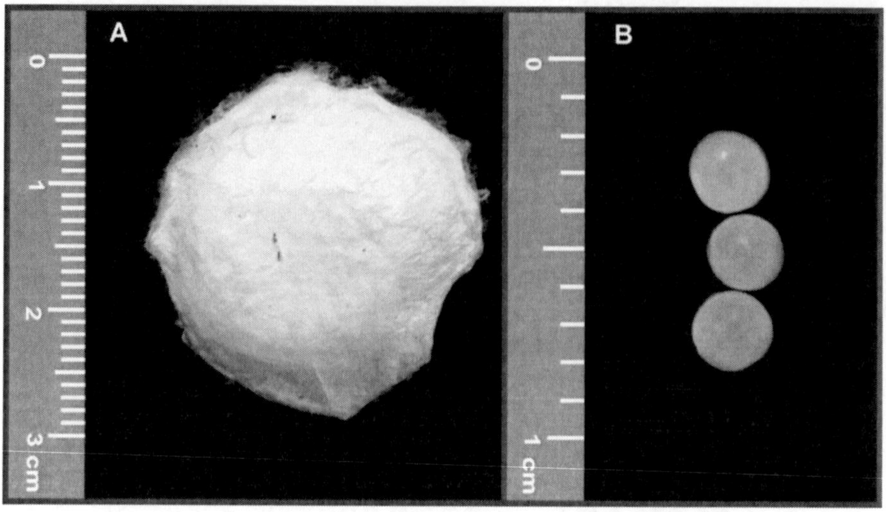

Figure 9. Photograph of ootheca and 5-day laid eggs of *P. pythagoricus*.

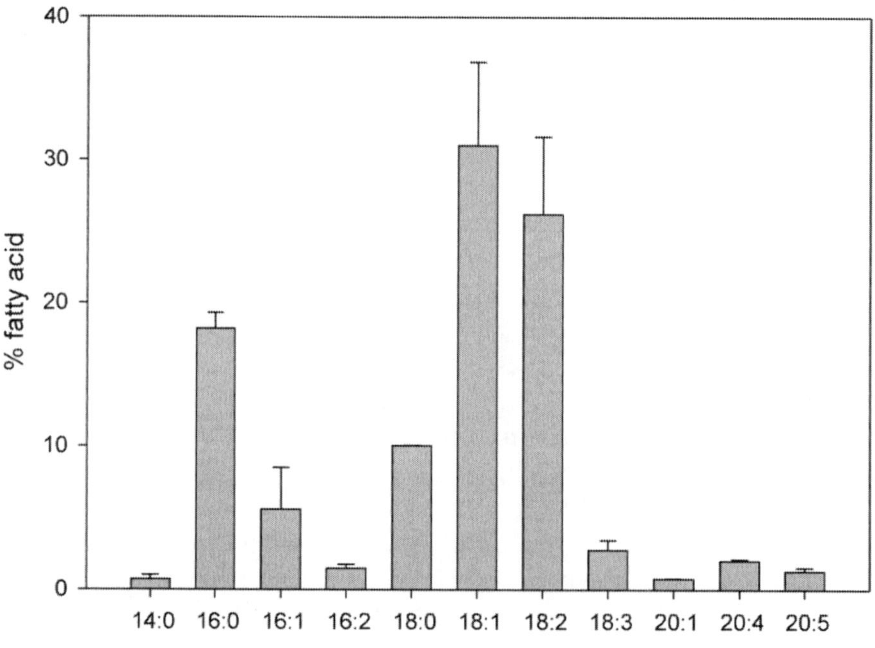

Figure 10. Fatty acid composition of LV1 (HDL) of eggs of *P. pythagoricus*. Data on fatty acid classes are expressed as weight percentage and quantified by GLC. Values are the mean +/- 1 SD of three independent analyses.

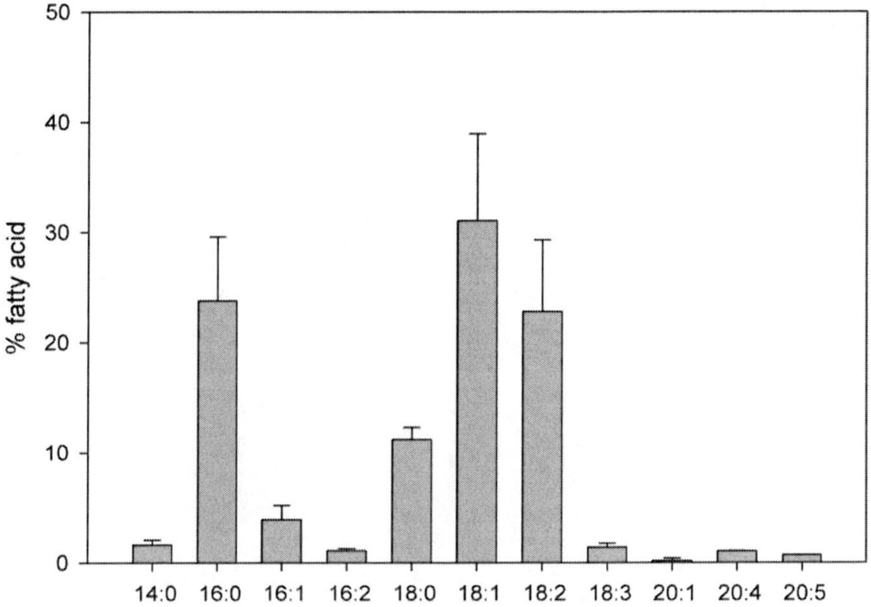

Figure 11. Fatty acid composition of LV2 (VHDL) of eggs of *P. pythagoricus.* Data on fatty acid classes are expressed as weight percentage and quantified by GLC. Values are the mean +/- 1 SD of three independent analyses.

In the Order Araneae a putative apovitellin in ovary and eggs of the spider *Tegenaria atricca* (Therafosidae) has already been described (Pourie and Trabalon, 2003). In the spider *P. pythagoricus* two lipovitellins named LV1 and LV2 have also been isolated from cytosol in eggs (Figure 9) and characterized (Laino et al., 2011). LV1 has 49 % lipids and density that corresponds to a high density lipoprotein (HDL), but LV2 has less lipid content, and it corresponds to a very high lipoprotein (VHDL). Of total lipids associated to lipoproteins, 91.5% are present in LV1, and only 8.4% are in LV2.

The major neutral lipid present in both particles was esterified colesterol (16% and 24%), triacylglycerols (8 and 9.5%) and diacylglycerols (3 and 4.3%) were also detected (Laino et al., 2011). Concerning structural lipids, phospholipids were the most abundant, 57% in LV1 and 47% in LV2 which were represented by phosphatidyl ethanolamine (PE) and PC (Table 3). It is evident that cytosol could have other energy lipids as TAG that can be found in droplets scattered throughout the cytoplasm.

The major fatty acid was 18:1 with 31% in both LVs, 18:2 with 26 and 22% ,16:0 with 18% and 23% in LV1 and LV2, respectively (Figs. 10 and 11) (Laino et al., 2011). Further studies are needed to come to a conclusion dealing with lipid content and fatty acid composition in spider eggs.

Molecular weights of lipoproteins were measured by native electrophoresis. LV1 showed 550 kDa molecular weight and LV2 evidenced 3 bands of 571, 430, 257 kDa (Figure 12). Under denaturing conditions LV1showed two main apolipoproteins of 64 and 25 kDa. In LV2 native bands were composed of the same subunits of 181, 67 y 60 kDa, but with different proportions (Table 3).

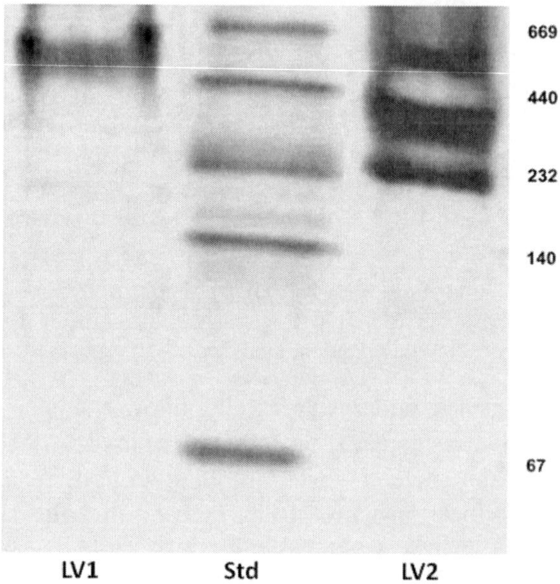

Figure 12. Native gel electrophoresis (4-23% acrylamide gradient slab) of LV1 and LV2 of *P. pythagoricus.* The proteins were visualized by Coomasie Brillant Blue.

P. pythagoricus lipovitellin, observed by electron microscopy using negatively stained preparations, appeared with spheroidal morphology (Figure 13) which was similar to insect plasma HDL (Laino et al., 2011) with a diameter of 20 nm in LV1 and 27.3 nm in LV2 (Laino et al., 2011). In contrast, crustacean HDL was reported to be discoidal in lobster and crab (Spaziani et al., 1986; Lee and Puppione, 1978). For crustaceans, these authors determined that plasma HDL did not have cholesterol esters but small amounts of triacilglycerides. Lipovitellin showed spheroidal morphology in *P. pythagoricus* due to the probable presence of neutral lipids.

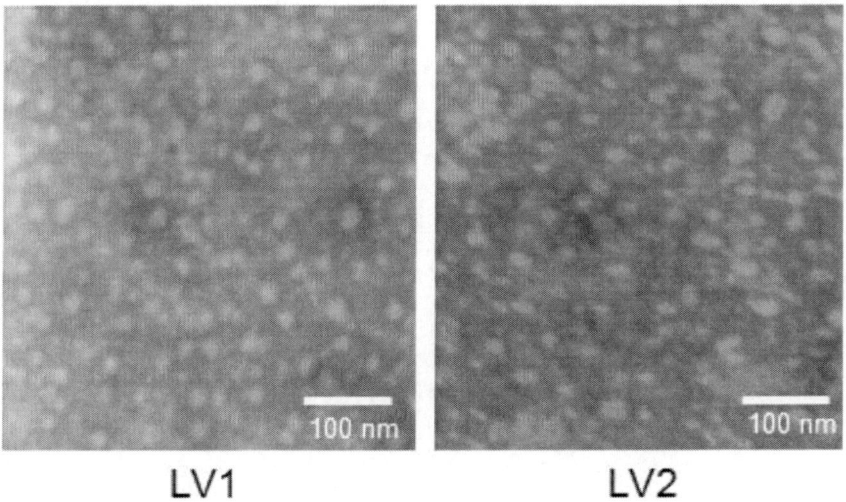

Figure 13. Electron micrographs of negatively stained of LV1 and LV2 of eggs from *P. pythagoricus*.

Table 3. Lipovitellins

Species Lipoproteins	Density gr/ml	% lipids	% proteins	% carbo-hydrates	Amount of lipids transported %	Lipid clases %	Particle mass kDa	Apolipopro-teins MW kDa
P. pythagoricus LV1 (HDL) (5)	1.16	49.3	49.3	1.4	91.5	PE 25 PC 23 EE 16	550	64 25
P. pythagoricus LV2 (VHDL) (5)	1.23	9.7	89.6	0.7	8.5	EE 24 PE 25 PC 24	571 400 257	181 67 60

Reference in table 1.

Due to the fact that LVs are not related to hemolymph lipoproteins as demonstrated by assays done with antibodies, it is likely that specific mechanisms of reproduction could produce these glycolipoproteins. Glycosylation of those lipoproteins closely related with reproduction has been reported in several species of Acari, (Cabrera et al., 2009), but small amounts of carbohydrates in lipovitellins of *P. pythagoricus* (1.4% and 0.7% in LV1 and LV2 respectively) were found.

Arachnids are a large group which showed be further investigated in order to gather more data about lipovitellins as energy source and those mechanisms involved in embryogenesis.

Acknowledgments

This work was supported by grants from Consejo Nacional de Investigaciones Científicas y Técnicas (CONICET-PIP no. 075) and Agencia Nacional de Promoción Científica y Tecnológica (PICT no. 2010-1270), Argentina. F.G. and M.C. are members of CONICET, Argentina. A.L. is recipient of a fellowship from CONICET, Argentina. We are grateful to Francisco Giambelluca for the photograph, Norma Tedesco for her English translation assistance and Mario R. Ramos for the figures.

References

Aparicio, G.M., Bonilla-Hurtado, V.A., Vargas, R.A. (2010). Caracterización mecánica y térmica del biopolímero hilo de araña *Nephila clavipes* como posible material para prótesis en hernia inguinal. *Rev. Col. de Física.* 42 [2], 176-180.

Armendano, A., González, A. (2010). Comunidad de arañas (Arachnida, Araneae) del cultivo de alfalfa (*Medicago sativa*) en Buenos Aires, Argentina. *Rev. Bio. Trop.* 58, 757-767.

Baird, S., Kelly, S.M., Price, N.C., Jaenicke, E., Meesters, C., Nillius, D., Decker, H., and Nairn, J. (2007). Hemocyanin conformational changes associated with SDS-induced phenol oxidase activation. *Biochim Biophys Acta.* 1774, 1380-1394.

Baker, M.E. (1988). Invertebrate vitellogenin is homologous to human von Willebrand factor. *Biochem. J.* 256, 1059-1061.

Beraud, E., Chandy, K.G. (2011). Therapeutic potential of peptide toxins that target ion channels. *Inflamm. Alergy Drug Target.* 10, 322-342.

Bianchi, A.G.de., Capurro, M.L., Marinotti, O. (1987). Lipohonn in the larval and adult stages of *Musca domestica. Arch. insect Biochem. Physiol.* 6, 39-48.

Boctor, F.N., Kamel, M.Y. (1976). Purification and characterization of two lipovitellins fro eggs of the tick, *Dermacentor andersoni*. *Insect Biochem.* 6, 233-240.

Boldbaatar, D., Umemiya-Shirafuji, R., Liao, M., Tanaka, T., Xuan, X., Fujisaki, K. (2010). Multiple vitellogenins from the *Haemaphysalis longicornis* tick are crucial for ovarian development . *J. Insect Physiol.* 56(11),1587-98.

Bridges, C.R. (2001). Modulation of haemocyanin oxygen affinity: properties and physiological implications in a changing world. *J. Exp. Biol.* 204, 1021-1032.

Cabrera, A.R., Donohue, K.V., Khalil, S.M.S., Sonenshine, D.E., Roe, R.M. (2009). Characterization of vitellin protein in the twospotted spider mite, *Tetranychus urticae* (Acari: Tetranychidae). *J. Insect Physiol.* 55, 655-661.

Cheli, G., Armendana, A., González, A. (2011) Feeding preferences of the spider *Misumenops pallidus* (Araneae: Thomisidae) on potential prey insects from alfalfa crops. *Rev.Bio.Trop.* 54(2), 505-513.

Chen, L., Jiang, H., Zhou, Z., Li, K., Li, K., Deng, G.Y., Liu, Z. (2004). Purification of vitellin from the ovary of Chinese mitten-handed crab (*Eriocheir sinensis*) and development of an antivitellin ELISA. *Comp Biochem. Physiol B.* 138, 305-311.

Chino, H. (1997). Physiological significance of lipid transport by lipophorin for long-distance flight in insects. *Comp. Biochem. Physiol.* B. 117, 455-461.

Chinzei, Y., Chino, H., Takahashi, K. (1983). Purification and propieties of vitellogenin and vitellin from a tick, *Ornithodoros moubata* . *J. Comp. Physiol.* B 152, 13-21.

Cunningham, M., García, F., Pollero, R.J. (2007). Arachnid lipoproteins: Comparative aspects. *Comp. Biochem. Physiol.* C. 146, 79-87.

Cunningham, M., Pollero, R.J. (1996). Characterization of lipoprotein fractions with high content of hemocyanin in the hemolymphatic plasma of *Polybetes pythagoricus*. *J. Exp. Zool.* 274, 275-280.

Cunningham, M., Pollero, R.J., Gonzalez, A. (1994). Lipid circulation in spiders. Transport of phospholipids, free acids and triacylglycerols as the major lipid classes by a high-density lipoprotein fraction isolated from plasma of *Polybetes pythagoricus*. *Comp. Biochem. Physiol.* B. 109, 333-338.

Cunningham, M.L., Gomez, C., Pollero, R.J. (1999). Lipid binding capacity of spider hemocyanin. *J. Exp. Zool.* 284, 368-373.

Cunningham, M.L., Gonzalez, A., Pollero, R.J. (2000). Characterization of lipoproteins isolated from the hemolymph of the spider *Latrodectus mirabilis*. *J. Arachnol.* 28, 49-55.

Decker, H., Jaenicke, E. (2004). Recent findings on phenoloxidase activity and antimicrobial activity of hemocyanins. *Dev. Comp. Immunol.* 28 (7-8), 673-687.

Decker, H., Rimke, T. (1998). Tarantula hemocyanin shows phenoloxidase activity. *J. Biol. Chem.* 273, 25889-25892.

Dhadialla, T.S. (1986). Purification and some biochemical propierties of vitellins from *Rhipicephalus appendiculatus* eggs and their use as antigens to induce type II immune resistance in rabits. *J. Cell. Biochem.* 10, 77.

Dhadialla, T.S., Raikhel, A.S. (1990). Biosynthesis of mosquito vitellogenin. *J. Biol. Chem.* 256, 9924-9923.

Dolaska-Angelova, P., Dolashki, A., Savvides, S.N., Hristova, R., Van Beeumen, J., Voelter, W., Devreese, B., Weser, U., di Muro, P., Salvato, B., Stevanovic, S. (2005). Structure of hemocyanin subunit CaeSS2 of the crustacean mediterranean crab *Carcinus aestuarii*. *J.Biochem.* 138, 303-312.

García, F., Cunningham, M., Soulages, J.L., Garda, H.A., Pollero, R.J. (2006). Structural characterization of the lipovitellin from the shrimp *Macrobrachium borellii*. *Comp. Biochem. Physiol. B.* 145, 365-370.

González, A., Lilgesthröm, G., Minervino, E., Castro, D., González, S., Armendano, A. (2009). Predation by *Misumenops pallidus* (Araneae: Thomisidae) on insects pests of soybeancultures in Buenos Aires Province, Argentina. *J. of Arach.* 37, 282-286.

Hall, M., van Heusden, M.C., Söderhäll, K. (1995). Identification of the major lipoproteins in crayfish hemolimph as proteins involved in immune recognition and clotting. *Biochem. Bio. Res. Co.* 216, 939-946.

Haunerland, N.H., Bowers, W.S. (1987). Lipoproteins in the hemolimph of the tarantula, *Eurypelma californicum*. *Comp. Biochem. Physiol. B* 86, 571-574.

Haunerland, N.H., Bowers, W.S. (1989). Comparative studies on arthropod lipoproteins. *Comp. Biochem. Physiol. B* 92, 137-141.

Heras, H., Pollero, R.J. (1990). Occurrence of plasma lipoproteins in octopods. Partial characterization and interorgan transport of lipids. *J. Exp. Mar. Biol. Ecol.* 140, 29-38.

Heras, H., Pollero, R.J. (1992). Hemocyanin as an apolipoprotein in the hemolymph of the cephalopod *Octopus tehuelchus*. *Biochim. Biophys. Acta.* 1125, 245-250.

Hoffmann, A. (2003). El maravilloso mundo de los arácnidos. Fondo de cultura economica. Consejo nacional de Ciencia y Tecnología, México D.F.

Jaenicke, E., Decker, H. (2004). Functional changes in the family of type 3 copper proteins in evolution. *Chem.Biol.Chem.* 5, 163-176.

James, A.M., Oliver, J.H., Jr. (1997). Purification and partial characterization of vitellin from the black-legged tick, *Ixodes scapularis. Insect Biochem. Molec. Biol.* 27, 639-649.

Kawazoe, I., Jasmani, S., Shih, T., Suzuki, Y., Aida, K. (2000). Purification and characterization of vitellin from the ovary of kuruma prawn, *Penaeus japonicus. Fisheries Sci.* 66, 390-396.

Kawooya, J.K., Keim, P.S., Ryan, R.O., Shapiro, J.P., Samaraweera, P., Law, J.H. (1984). Insect apolipophorin III: Purifícation and properties. *J Biol Chem.* 259, 10733-10737.

Laino, A., Cunningham, M., Heras, H., and García, F. (2011). Isolation and characterization of two vitellins from eggs of the spider *Polybetes pythagoricus* (Araneae: Sparassidae). *Comp. Biochem. Physiol. B.* 158, 142-148.

Laino, A., Cunningham, M.L., Garcia, F., Heras, H. (2009). First insight into the lipid uptake, storage and mobilization in arachnids: role of midgut diverticula and lipoproteins. *J. Insect Physiol.* 55, 1118-1124.

Lee, R.F., Puppione, D.L. (1978). Serum lipoproteins in the spiny lobster, *Panulirus interruptus. Comp. Biochem. Physiol.* B 59, 239-243.

Lubzens, E., Ravid, T., Khayat, M., Daube, N., Tietz, A. (1997). Isolation and characterization of the high-density lipoproteins from the hemolymph and ovary of the penaeid shrimp *Penaeus semisulcatus* (de Haan): apoproteins and lipids. *J. Exp. Zool.* 278, 339-348.

Markl, J. (1986). Evolution and function of structurally diverse subunits in the respiratory protein hemocyanin from arthropods. *Biol. Bull.* 171, 90-1145.

Numata, K., Reagan, M.R., Goldstain, R.H., Rosenblatt, M., and Kaplan, D.L. (2011). Spider Silk-Based Gene Carriers for Tumor Cell-Specific Delivery. *Bioconjug Chem* . 22(8):1605-10.

Pourie, G., Trabalon, M. (2003). The role of 20-hydroxyecdysone on the control of spider vitellogenesis. *Gen. Comp. Endocrinol.* 131, 250-257.

Rosell, R., Coons, L.B. (1992). The role of the fat body, midgut and ovary in vitellogenin production and vitellogenesis in the female tick, *Dermacentor variabilis. Int. J. Parasitol.* 22, 341-349.

Ryan, R.O., Schmidt, J.O., Law, J.H. (1984). Chernical and immunological properties of lipophorins from seven insect orders. *Arch Insect Biochem Physiol.* 1, 375-383.

Sahni, V., Blackledge, T.a., and Dhinojwala, A. (2010). Viscoelastic solids explain spider web stickiness. *Nature comm.* 1(2), 1-4.

Salerno, A.P., Dansa-Petretski, M., Silva-Neto, M.A.C., Coelho, H.S.L., Masuda, H. (2002). *Rhodnius prolixus* vitellin is composed of three different populations: comparison with vitellogenin. *Insect Biochem. Molec.* 32, 709-717.

Scanu, A.M., Edelstein, C., Keim, P. (1975). Serum lipoproteins, en the plasma protein. structure, function and genetic contrrol. Academic Press, New York.

Schacht, K., Scheibel, T. (2011). Controlled hydrogel formation of a recombinant spider silk protein. *Biomacromolecules.* 12(7), 2488-2495.

Schartau, W., Leidescher, T. (1983). Composition of the hemolimph of the tarantula *Eurypelma californicum. J. Com. Physiol.* 152, 73-77.

Schriefer, M.E. (1991). Vitellogenesis in *Hyalomma dromedarii* (Acari: Ixodidae): A model for analysis of endocrine regulation in ixodid ticks. Old Dominion University.

Shanbaky, N.M., Main, A.J., El-Said, A., Helmy, N. (1990). Vitellogenic and non vitellogenic proteins in haemolimph, ovaries, and eggs of *Argas (Persicargas) arboreus. Exp. Parasitol.* 27, 986-992.

Shapiro, J.P., Law, J.H., Wells, M.A. (1988). Lipid transport in insects. *Annu. Rev. Entomol.* 33, 297-318.

Spaziani, E., Havel, R.J., Hamilton, R.L., Hardman, D.A., Stoudemire, J.B., Watson, R.D. (1986). Properties of serum high-density lipoproteins in the crab, *Cancer antennarius* Stimpson. *Comp. Biochem and Physiol. B* 85 (2), 307-314.

Steward, G.A., Fisher, W.F. (1985). Lipoproteins from the house dust mite *Dermatophagoides pteronyssinus* and the mangq mites *Psoroptes cuniculi* and *Psoroptes ovis. Comp.Biochem.Physiol.B.* 81, 915-922. 1985.

Stratakis, E., Fragkiadakis, G., Tentes, I. (1993). Purification and properties of the fatty acid-binding VHDL from the hemolymph of the spider *Eurypelma californicum. J. Exp. Zool.* 267, 483-492.

Tatchell, R.J. (1971). Electrophoretic studies on the proteins of the haemolimph, saliva and eggs of the cattle tick, *Dermacentor variabilis* . *Insect Biochem.* 1, 47-55.

Terwilliger, N.B. (1998). Functional adaptations of oxygen-transport proteins. *J. Exp. Biol.* 201, 1085-1098.

Tufail, M., Takeda, M. (2008). Molecular characteristics of insect vitellogenins. *J. Insect Physiol.* 54, 1447-1458.

Van Bruggen, E.F.J. (1980). Hemocyanin: the mystery of blue blood. *Trends in Biol. Sci.* 7, 185-188.

Voit, R., Feldmaier-Fuchs, G., Schweikardt, T., Decker, H., Burmester, T. (2000). Complete sequence of the 24-mer hemocyanin of the tarantula *Eurypelma californicum.* Structure and intramolecular evolution of the subunits. *J. Biol Chem.* 275, 39339-39344.

Walker, A., Ando, S., Smith, G.D., Lee, R.F. (2006). The utilization of lipovitellin during blue crab (*Callinectes sapidus*) embryogenesis. *Comp. Biochem. Physiol.* B 143, 201-208.

Wolfgang, N. (2012). The species referred to as *Eurypelma californicum* (Theraphosidae) in more than 100 publications is likely to be *Aphonophelma hentzi. J. Arachnol.* 40, 128-130.

Yepiz-Plascencia, G., Vargas-Albores, F., Higuera-Ciapara, I. (2000). Penaeid shrimp hemolymph lipoproteins. *Aquaculture* 191, 177-189.

Zatta, P. (1981). Protein-lipid interactions in *Carcinus maenas* (Crustacea) hemocyanin. *Comp. Biochem. Physiol.* B 69, 731-735.

In: Spiders
Editor: Marq Santerre

ISBN: 978-1-62618-517-3

Chapter 4

STUDY OF THE DISCRIMINATE INFORMATION ON SPIDER WEBS

Jaime R. Ticay-Rivas, Carlos M. Travieso, Marcos del Pozo-Baños, Jesús B. Alonso and William G. Eberhard
University of Las Palmas de Gran Canaria, Spain

ABSTRACT

This work presents the demonstration of the biological hypothesis about the localization of the main information of the web spider, in order to apply it for identifying of spider species. The question is the analysis of the whole the web spider or only the center of the spider web, as the hypothesis says. After our experiments with our database, applying a pre-processing stage, enhanced the spider web images and using different kinds of parameterization methods for classifying it with support vector machines; this work have demonstrated the hypothesis is true, and the most important information is located on the center of the spider web.

INTRODUCTION

The biodiversity conservation is a priority issue that must be urgently attended (Unesco, 2012) since it is vital to all living forms on the planet. It is recognized by the scientific community that species research is fundamental in order to design and develop the right methodologies to understand and protect the biodiversity. Therefore, the biodiversity conservation is one of the principal topics for researchers (Sytnik, K.M., 2010).

There is a close relation between the climate change and the biodiversity. On one side, the climate change affects the biodiversity and on the other side, the biodiversity contributes to both climate-change mitigation and adaptation by using the ecosystem services that it supports. Thus, is necessary a sustainable management of the biodiversity to confront the current climate changes.

Moreover, thousand of plants and animal species are currently threatened with extinction because of habitats destruction. Overexploitation, limitations of the areas where these species are distributed and meanly, the lack of knowledge on the biodiversity conservation, bring as a fatal consequence the decrement of the population of the species. As a general rule, it is assumed that if the original area of a habitat is reduced to its tenth part, it will lose the half of the species contained. These consequences are more striking in the rainforests due to increased rates of deforestation.

The development of biodiversity policy depends on demographic, biological and ecological knowledge. Assessment of abundance, survival and life-history characteristics are available by mark and recapture in animal studies. In the flora research, the collecting is the usual method to obtain the samples for the subsequent analysis. Since this method provide a large databases, is very hard to exploit the full potential that can be achieved due the unmanageable task of perform a pattern recognition in large data sets.

The researchers are working hard to comprehend the existing biodiversity and they are using all the methods and tools available to accomplish that aim. The automatic recognition is a recent powerful technique for help the ecological research methodologies to be more effective and efficient. The traditional approaches mainly depend on human expertise, which is hence both times consuming and costly.

The spiders are plentiful; ecologically essentials (Johnston, 2000), (Peterson, Osborne and Taylor, 1989), (Cardoso, Arnedo, Triantis and Borges, 2010) and they present the ability to react to environmental changes and impacts of the human being on the nature (Cardoso et al., 2010), (Finch, Blick, and Schuldt, 2008). Despite these facts, the spiders have been one of the most unexplored groups in conservation biology (Carvalho, Cardoso, Crespo, Henriques, Carvalho, and Gomes, 2011). In order to provide new knowledge on spiders' nature, this chapter presents the results of the study of the spider species recognition by using the information contained in their cobwebs. This work applies the biologists' theory that defines the center of the cobwebs as principal region of information. This conclusion is built on the heuristic experimentation and the expertise of the biologist specialized on spiders.

BACKGROUND

In the state of art, it can be found two works where authors emphasized their study on the way spiders build their webs for species identification (Eberhard, 1982), (Eberhard, 1990). A model which described spider behavior by simulating how specific spider specie builds its web it was proposed in (Eberhard, 1969). In (Suresh, 2000) was recorded a controlled scenario for further spatio-temporal analysis.

INFORMATION OF THE WEB SPIDER

The goal of this work is to determine where the relevant information of the spider web is located in order to use it for the identification of the spider species. In biological publications has been demonstrated that, this knowledge can be acquired after years of experience. The aim of this work is to implement an automatic approach to this process, which gives a quantitative and objective answer, using the parameterization algorithm and artificial intelligence. The Biologists' hypothesis is based on the localization of the main information to identify spider species from its spider web. They have defined the center of the spider web as the main source of discriminative information. We have tried getting that information in order to evaluate this affirmation. For it, the initial point is our database. A pre-processing stage is applied in order to remove the noises and interferences, until to get the spider web. After this, different kinds of parameterization algorithms have been applied.

SPIDER DATABASE

The database is formed by spider web images of four species named Allocyclosa, Anapisona Simoni, Micrathena Duodecimspinosa and Zosis

Geniculata. Each class has respectively 28, 41, 39 and 49 images, which makes a total of 150 images. Some examples can be seen in Figure 1. The spider webs images were taken in both controlled and uncontrolled environments, thus, the lightness condition and background differ between them.

In order to develop this work, the images of each specie were assigned to a specific class. To Class 1 (C1) were assigned the images that correspond to the specie named Allocyclosa. These images were taken in a natural night time environment. The flash of the camera in conjunction with dark background enhanced the spider webs, resulting in a set of images with good quality for the processing step.

The Anapisona Simoni images were assigned to Class 2 (C2). Theses spider webs were built in a controlled environment, with the singularity that they were built as tents, causing overlapping thread and light reflections. This made the processing stage far more more complex than in C1.

Class 3 (C3) is composed by Micrathena Duodimspinosa. These images were also taken in a natural environment during the day. This scenario, with the presence of the sun light and natural elements such as leaves and tree branches, required a more complex treatment as well.

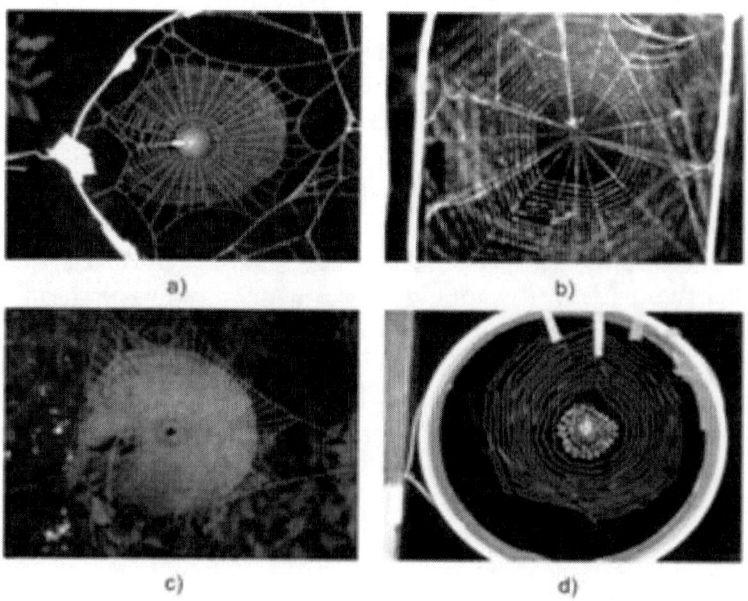

Figure 1. Spider web samples of the four species: a) Allocyclosa. b) Anapisona Simoni. c) Micrathena Duodimspinosa. d) Zosis Geniculata.

Table 1. Technical features of the database

Class	Number of samples	Size (pixels)	Bits number (color/gray)
1	28	1024x768	24 (True color)
2	41	2240x1488	8 (Gray scale)
3	39	2240x1488	24 (True color)
4	42	2216x2112	24 (True color)

Finally, the Zosis Geniculata images were assigned to the Class 4 (C4). Again, they were taken in a controlled environment, allowing the capture of images with black background and uniform light.

The technique characteristics of images from each class are summarized in Table 1.

PREPROCESSING

The spider web images used in this work were taken in both controlled and uncontrolled environments as can be seen in Figure 1. Therefore, the image treatment was fundamental in order to isolate the spider webs and remove effects of background in the system.

To achieve this goal, image processing techniques were applied to isolate the spider webs from light reflections and elements of the background of the image such as leaves or tree branches. After this step, a new normalized database was obtained.

1. Spider Web Selection

The images that compose the database were not taken for this type of research, therefore, there are information that will introduce noise to the system.

This noise corresponds to any external element of the spider webs. To solve this problem, the region of interest of the image was manually selected in order to ensure that only the potential source of information is considered for the study. This is represented in Figure 2.

Once the spider web has been select, an adjustment of the proportional ratio was necessary in order to obtain a proportional square image. This will be explained in detail in the following section.

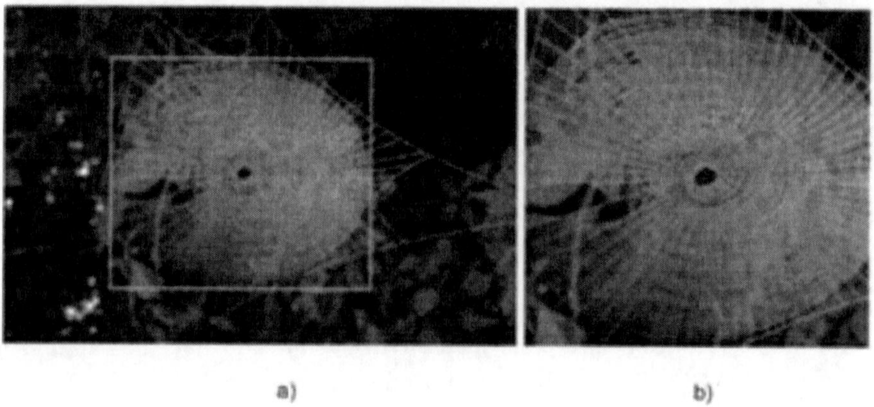

Figure 2. Spider web Selection.

2. Image Contrast

An increase of the color contrast was applied as first step in order to enhance the contour of cobweb's threads.

In order to obtain a better visual interpretation and prepare the data for the next preprocessing step a spacial filtering was applied, this allow to enhance or attenuate details in the image analysed. By using this filtering, the value of each pixel is modified according to neighbors' values, transforming the original gray levels so that they become more similar or different to the corresponding neighboring pixels.

In general, the convolution of a image with dimensions with a mask is given by equation 1:

$$g(x,y)=\sum_{s=-a}^{a}\sum_{t=-b}^{b}f(x+s,y+t)h(s,t) \tag{1}$$

where are the pixel value's of the selected block, are the mask coefficients and is the filtered image. The block dimension is defined by and $n=2b+1$. In Figure 3 it can be observed the effect of applying contrast enhancement filtering to a gray image a), the result obtained in b) shows the contour of the threads of the cobwebs more shaped and uniform.

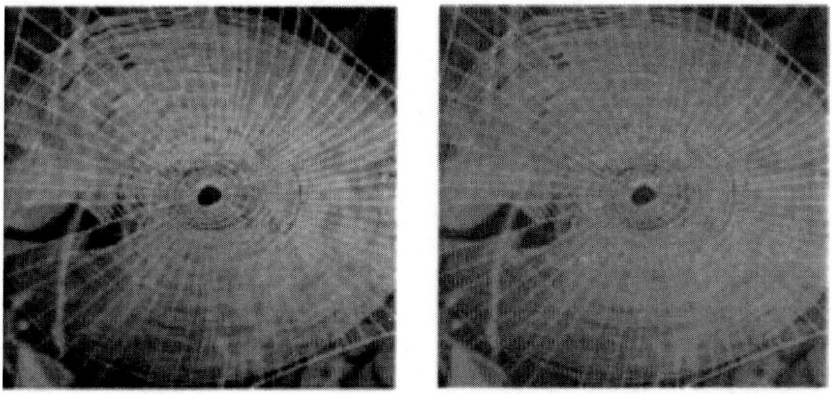

a) b)

Figure 3. a) Original Image b) Contrast enhanced.

3. Image Binarization

The binarization process transforms the image to a black and white format in a way that it does not change the essential properties of the image. Equation 2 defines the binarization process, where is the original image and the obtained image:

$$g(x,y)=\begin{cases}0 & f(x,y)<threshold\\ 1 & f(x,y)\geq threshold\end{cases} \tag{2}$$

This threshold is computed by using the very know Otsu's method (Otsu, 1979), which assigns the membership of each pixel to a determined group by computing the optimal value from which carrying out that assignment.

3. Image Denoising

Once the image has been binarized, a denoising process was used aiming to eliminate any irrelevant information. To achieve this goal, two specific techniques were applied: Wiener Filtering and Morphological Operations.

The Wiener Filtering apply a spatial filtering using statistical methods in order to reduce noise and smooth shapes. It gradually smooths the image by changing the areas where the noise is very apparent, but keeping the areas where the details are present and the noise is less apparent. The Wiener filter is adapted to the local image variance. The Wiener filter compute the local mean and the variance around each pixel in the image a;

$$\mu = \frac{1}{NM} \sum_{\eta_1, \eta_2 \in \eta} a(\eta_1, \eta_2) \tag{3}$$

$$\sigma^2 = \frac{1}{NM} \sum_{\eta_1, \eta_2 \in \eta} a^2(\eta_1, \eta_2) - \mu^2 \tag{4}$$

where is defined as the local neighborhood for each pixel N × M in the image a. An block has been chosen by euristics, i.e. this was the configuration that provided the best visual effect. Then, the image is filtered using these estimates, where is the resulting image. If the noise variance is not given, the filter uses the average of all the local estimated variances.

$$b(n_1, b_2) = \mu + \frac{\sigma^2 - v^2}{\sigma^2}(a^2(\eta_1, \eta_2) - \mu) \tag{5}$$

On the other hand, morphological operations are those transformations that modify the structure or shape of the objects in the image based on the their geometry and shape, simplifying the images. These techniques can be used to denoise an image, for feature extraction or processing specific regions.

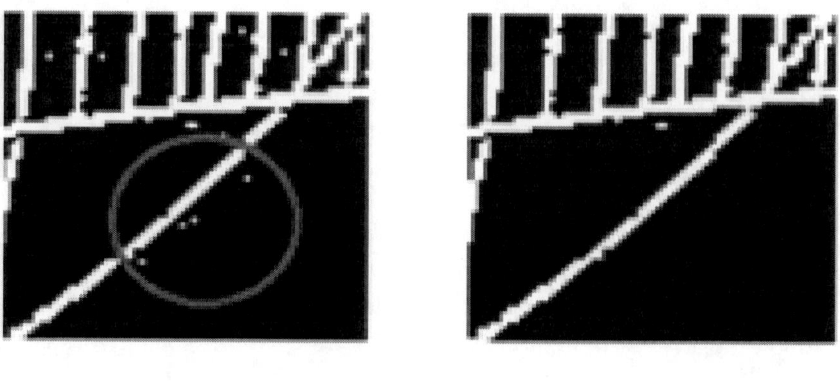

a) b)

Figure 4. Example of applying morphological operations: elimination of isolate pixels.

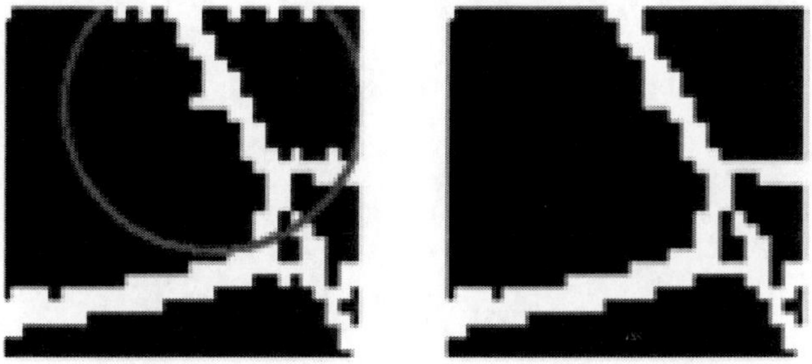

a) b)

Figure 5. Example of applying morphological operations: smoothing of contour.

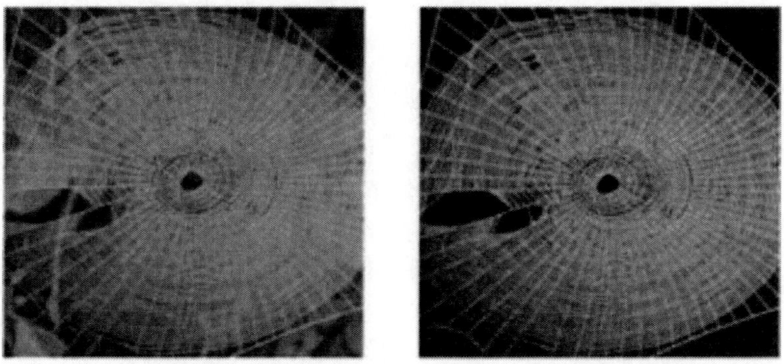

a) b)

Figure 6. Image after denoising. a) Original image b) Image binarized.

An illustrative example of these operations is shown in Figures 4 and Figure 5, where noise and projections (in the inner circle) are removed, obtaining more uniform boundaries. The resulting image after image denoise can be observed in Figure 6.

4. Center of the Spiderwebs

Finally, the center of the spiderwebs was selected in order to verify where the most of the information is contained.

Thus, once the images were preprocessed this center area was selected to conform the second database. Figures 7 and 8 show the center of the spiderwebs for each specie.

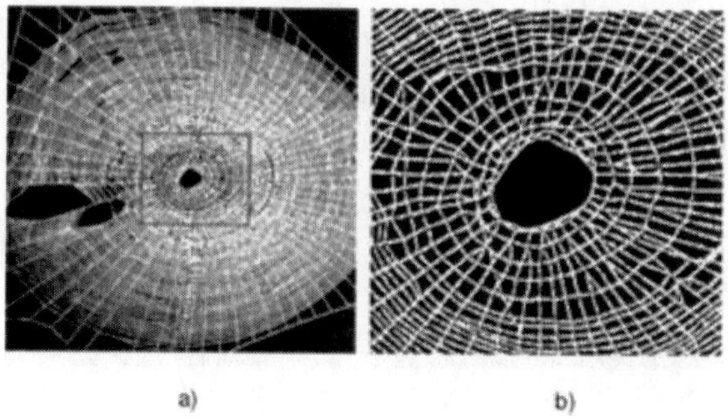

Figure 7. Selecting the center of the spiderwebs.

Figure 8. Resulting center of each specie.

FEATURES EXTRACTORS

This work proposes an approach to face the problem of spider specie recognition by using the information that underlie in their cobwebs. In order to achieve this aim, it is presented a comprehensive study on the reliability of the most used features extractors: Principal Component Analysis (PCA) (Pearson, 1901) , Independent Component Analysis (ICA), (Comon, 1994) , Discrete Cosine Transform (DCT) (Ahmed, Natarajan, and Rao, 1974), Wavelet

Transform (DWT) (Akansu and Haddad, 1992) and Discriminative Common Vectors (DCV) (Travieso, Botella, Alonso, and Ferrer, 2009).

EVALUATION OF THE INFORMATION OF THE WEB SPIDER

The evaluation has been developed by Support Vector Machines (SVM) (Vapnik, 1995) The SVM is a method of structural risk minimization (SRM) derived from the statistical learning theory developed by Vapnik and Chervonenkis (Vapnik, 1995). It is enclosed in the group of supervised learning methods of pattern recognition, and it is used for classification and regression analysis.

Based on characteristic points called Support Vectors (SVs), the SVM uses an hyperplane or a set of hyperplanes to divide the space in zones enclosing a common class. Labeling these zones the system is able to identify the membership of a testing sample. The interesting aspect of the SVM is that it is able to do so even when the problem is not linearly separable. This is achieved by projecting the problem into a higher dimensional space where the classes are linearly separable. The projection is performed by an operator known as kernel, and this technique is called the kernel trick (Kevman, 2001), (Schölkopf and Smola, 2002). The use of hyperplanes to divide the space gives rise to margins.

In this work, the LS-SVM (Suykens, Van Gestel, De Brabanter, De Moore and Vandewalle, 2002) was used along with the Radial Basis Function kernel (RBF-kernel). The regularization parameter and the bandwidth of the RBF function were automatically optimized by the validation results obtained from 10 iterations of a Hold-Out cross-validation process (Arlot, 2010). Two samples from each class (from the training set) were used for testing and the remaining for training as we saw that the number of training samples has a big impact in the LS-SVM optimal parameters. Once the optimal parameters were found, they were used to retrain the LS-SVM using all available training samples.

The main experiment has been carried out was the localization of the information. The systems were configured in identification mode and the optimal size was compute in order to reach the maximum performance for the two next experiments. Once the optimal size was obtained, in the second experiment, the best configuration and the optimal parameters obtained in the previous experiment were used in verification mode. In order to obtain quantitative values about where the valid information is contained, all the experiments were performed using the dataset described, that corresponds to the center area of the cobwebs dataset obtained.

There were designed eleven systems. The first five of them corresponds to the implementation of each feature extractor technique. The remaining correspond to a serie of fusion systems proposend in this work. For all the experiments, the systems were designed by applying a supervised learning strategy and using the first M features obtained from the PCA, ICA, DCT projection of the spider webs images and the outcome of the DWT and DCV transformation of the spider webs images as inputs for a RBF-kernel LS-SVM with regularization and kernel parameters. The former parameter (the number of features) was varied during experimentation, while the later two parameters (the regularization and the kernel parameters) were automatically optimized by iteration using validation results.

To obtain more reliable performance, the available samples were divided into training and test sets by using the K-Folds cross-validation and Hold-Out cross-validation techniques (Arlot, 2010), so that the system was trained and tested with totally different samples. In particular for the final results, experiments with K equal 3, 5, 7, and 10 were run. The percent of training samples for Hold-Out cross-validation technique was 50, 40, 30, 20, 10 percent respectively. It is worth it to mention that the training and testing sets were computed for each class individually, having into account that each class has different number of samples.

The main experiment of this work is to locate the most discriminative information for two configuration, the whole spider web and the center spider web. It was found in order to obtain the maximum performance possible in the final systems.

This computation was carried out by varying the size of the spiderweb images using each feature extractors proposed and then, the success rate reached for each size fixed was used to determine the optimal size. This procedure was applied both the whole spider web as the center of the spider web databases. This step is necessary because it help to determine in which size the features extractor can reach its maximum performance and moreover, ensure that the training time will be reasonable.

The experiment was carry out using the feature extractors techniques mentioned and with the system configured in identification mode. The results are shown as the average of 30 iterations of 50% Hold-Out cross-validation. The size was swept from 5×5 pixels to 50×50 pixels with steps of 5×5 pixels. Tables 2 and 3 shows the results of each system for each size of the whole spider web database.

Table 2. Results obtained for different sizes and systems for the whole spider webs

Success Rate (%) ± desv					
System	5×5	10×10	15×15	20×20	25×25
PCA	93.78% ± 2.65	95.14 ± 2.45	94.78% ± 2.48	94.78 2.15	94.19 2.44
Dmay	91.67% ± 3.12	94.23 ± 2.88	94.23% ± 2.58	94.41 2.92	94.41 2.43
Db1	92.48% ± 2.37	94.10 ± 3.04	94.19% ± 3.01	94.82 2.38	94.91 2.74
DCT	94.23% ± 2.60	93.42 ± 2.37	92.12% ± 2.38	90.68 1.89	90.68 2.76
ICA	91.08% ± 5.93	93.42 ± 2.53	93.38% ± 3.14	91.85 4.92	89.23 7.44
Bior	92.03% ± 1.92	94.05 ± 3.14	94.50% ± 2.43	94.73 2.47	94.55 2.44
DCV	87.84% ± 3.82	88.24 ± 3.39	88.87% ± 3.49	88.82 3.57	88.65 3.85

Table 3. Results obtained for different sizes and systems for the whole spider webs

Success Rate (%) ± desv					
System	30×30	35×35	40×40	45×45	50×50
PCA	93.87% 2.35	93.56 2.48	93.69% 2.93	93.11 2.78	93.24 2.66
Dmay	93.96% 3.47	94.05 ± 2.32	94.05% 2.42	93.96 1.76	94.60 2.77
Db1	94.68% 2.63	94.64 2.83	94.91% 2.37	94.86 2.42	94.41 4.13
DCT	90.45% 3.33	89.30 2.83	89.46% 2.64	89.10 2.95	89.32 3.04
ICA	88.38% 9.59	88.65 10.13	88.15% 10.20	88.65 9.27	87.34 9.33
Bior	94.55% 2.17	94.32 2.28	94.46% 2.47	94.19 2.38	93.51 3.66
DCV	88.65% 3.52	87.32 3.04	87.84% 3.37	87.79 3.50	87.43 3.59

Table 4. Results obtained for different sizes and system for the center of the spider webs

Success Rate (%) ± desv					
System	5×5	10×10	15×15	20×20	25×25
PCA	97.52% 1.81	99.41 0.85	99.32% 1.11	99.10 1.02	98.78 1.48
Dmay	98.29% 1.66	98.69 1.20	97.57% 2.14	97.39 1.91	95.95 2.49
Db1	95.32% 2.00	98.42 1.74	98.20% 1.64	98.51 1.30	98.65 1.12
DCT	95.32% 2.12	98.20 1.64	99.10% 1.25	98.47 1.54	98.60 1.30
ICA	97.03% 1.64	98.15 3.96	98.60% 3.90	97.16 2.47	97.75 3.04
Bior	95.95% 2.66	97.70 2.10	98.15% 1.76	98.11 2.55	98.56 1.62
DCV	95.23% 2.34	94.60 2.16	94.60% 1.88	94.28 2.34	94.05 2.50

Table 5. Results obtained for different sizes and systems for the center of the spider webs

Sucess Rate (%) ± desv					
System	30×30	35×35	40×40	45×45	50×50
PCA	98.29% 1.70	97.66 2.19	97.75% 1.79	97.39 1.62	97.30 1.67
Dmay	95.81% 2.49	95.63 2.18	95.81% 2.02	96.12 2.12	95.86 2.53
Db1	98.60% 1.30	98.24 2.56	98.11% 2.55	97.97 3.51	97.61 2.18
DCT	98.51% 1.20	98.02 1.58	97.43% 2.73	97.66 4.18	96.71 2.53
ICA	96.22% 5.34	96.17 4.60	95.59% 6.07	94.59 6.99	95.09 4.63
Bior	98.33% 1.26	98.11 2.57	98.56% 2.10	97.57 3.89	98.42 2.22
DCV	97.37% 1.81	93.83 2.06	93.92% 2.18	93.65 2.49	93.24 2.61

Table 6. Center spider web results Vs Whole spider web results

Sucess Rate (%) ± desv		
System	Center spider web (10×10 pixels)	Whole spider web (10×10 pixels)
PCA	99.41% 0.85	95.14 2.44
DCT	98.69% 1.20	93.42 2.37
Db1	98.42% 1.74	94.10 3.04
Bior	98.20% 1.64	94.05 3.14
ICA	98.15% 3.96	93.42 2.53
Dmey	97.71% 2.10	94.23 2.88
DCV	94.60% 2.16	88.24 3.39

The same methodology was applied to the the center of spider webs images. The Tables 4 and 5 shows the results obtained by using this dataset. As can be observed, the best results correspond to 10×10 pixels in all the systems, therefore, it was defined as the optimal size. Once the optimal size was found, the results allow to do the first approximation about which database present the best behaivor. The Table 6 show a comparition between the results obtained using the center of the spider web dabase vs the whole spider web database with the optimal size. The center of the spider web images seems to contain more information with a higher level of stability.

SOLUTIONS AND RECOMMENDATIONS

The systems developed were divided in three phases. The first phase involve the preprocessing of the images, where the center of the spider web images were extracted and the effects of the background were removed. In the second phase the resulting images were then transformed by using PCA, ICA, DCV, DWT and DCT. The optimal number of PCA, ICA and DCT coefficients M was found by euristics, while for DCV, its algorithm fixed this number to the number of classes -1. Moreover, the families Db1, Bior and Dmey were tested for the DWT. In the third phase, the resulting characteristics were classified using an LS-SVM. The regularization and kernel parameters were automatically optimized by the system dividing the training samples in training and validation sets and retraining the system with the optimal configuration using all available training data. The combination of these features extractor was proposed in order to improve the success rate reached with the individual systems.

As can be observed in tables 2 to 5, the optimal size of the images was 10×10 pixels, quite compressed, considering the spatial distribution of the threads in the spider webs. The results shows that the database formed by the center of the spider web compared to the database formed by the whole spider web was a better source of information as shows the table 6.

The results shows that the best feature extractor correspond to PCA-based system, reaching 99.41% of success rate in identification mode; and in general, the most important information is located in the center of the web spider.

CONCLUSION

This chapter has presented the results about the discriminate information on web spider to face the problem of spider recognition. In order to achive this aim, the most used features extractors were implemented and has been demostrated after a serie of experiments, that the main information from web spiders is located in its center. The work carried out has evaluated this hypothesis, which was kwown by the experts on spiders.

Another interesting point of this work, it is that this kind of study can be applied to other fields and so, evaluate other hypothesis.

ACKNOWLEDGMENTS

This work has been supported by Spanish Government, in particular by "Agencia Española de Cooperación Internacional para el Desarrollo" under funds from D/027406/09 for 2010, D/033858/10 for 2011 and A1/039089/11 for 2012.

Additional Reading

http://www.nhm.ac.uk/nature-online/biodiversity/

http://www.conservation.org/learn/biodiversity/Pages/overview.aspx

http://www.biometrics.org/

http://www.biometricsinfo.org/

Barham, P.J., Burghardt, T., and Sherley, R.B. (2010). Non-invasive Monitoring of Colonial Seabirds using Computer Vision: Scope and Limitations as Exemplified on African Penguins and Bank Cormorants. 1st World Seabird Conference (WSC2010), C4-1—11.

Eberhard, W. G. (1977). Aggressive Chemical Mimicry by a Bolas Spider. *Science*, 198 (4322): 1173–1175.

Eberhard, W. G. (1980). The Natural History and Behavior of the Bolas Spider, Mastophora dizzydeani sp. n. (Araneae). *Psyche,* 87 (3–4): 143–170.

Ethem Alpaydın (2004) Introduction to Machine Learning (Adaptive Computation and Machine Learning), MIT Press, ISBN 0-262-01211-1.

Ethem Alpaydın (2004) Introduction to Machine Learning (Adaptive Computation and Machine Learning), MIT Press, ISBN 0-262-01211-1.

Foelix, R. F. (1996). Biology of Spiders. 198 Madison Ave. NY, New York, 10016: Oxford University Press. p. 3. ISBN 0-19-509593-6.

George, L., and William, S. (2004). Artificial Intelligence: Structures and Strategies for Complex Problem Solving (5th Ed.). The Benjamin/Cummings Publishing Company, Inc. ISBN 0-8053-4780-1.

Haibo Yao; and Lei Tian (2003). A genetic-algorithm-based selective principal component analysis (GA-SPCA) method for high-dimensional data feature extraction. Geoscience and Remote Sensing, *IEEE Transactions,* 41(6), 1469- 1478.

Jing Hu, Si, J., Olson, B.P.,and Jiping He (2005). Feature detection in motor cortical spikes by principal component analysis. Neural Systems and Rehabilitation Engineering, *IEEE Transactions,* 13(3), 256-262.

Kays, W. R. , Ray, J.C., Gompper, M.E., and Zielinski, J.Z. (2007). Using Patterns in Track-Plate Footprints to Identify Individual Fishers. *The Journal of Wildlife Management,* 71 (3), 955-963.

Krijger, H., and Bangay, S., and Foster, G (2002). Individual Zebra Identification. - An investigation into image pre-processing, stripe extraction and pattern matching techniques.

Langley, P., Bowers, E.J., and Murray, A. (2010).Principal Component Analysis as a Tool for Analyzing Beat-to-Beat Changes in ECG Features: Application to ECG-Derived Respiration. Biomedical Engineering, *IEEE Transactions,* 57(4), 821-829.

Mallat, S. (1989). A theory for multiresolution signal decomposition: the wavelet representation. *IEEE Pattern Analysis and Machine Intelligence,* 11(7), 674-693.

Nan Liu, and Han Wang (2006). Feature Extraction with Genetic Algorithms Based Nonlinear Principal Component Analysis for Face Recognition," Pattern Recognition, *18th International Conference,* 3, 461-464.

Nilsson, N. (2010). The Quest for Artificial Intelligence: A History of Ideas and Achievements. New York: Cambridge University Press, ISBN 978-0-521-12293-1.

Qingfu Zhang, and Yiu Wing Leung (2000). A class of learning algorithms for principal component analysis and minor component analysis. Neural Networks, *IEEE Transactions,* 11(1), 200-204.

Roweis, S. T., and Saul, L. K. (2000). Nonlinear Dimensionality Reduction by Locally Linear Embedding. *Science,* 290 (5500), 2323–2326.

Score Fusion. Lane Department of Computer Science and Electrical Engineering, Department of Statistics West Virginia University.

Samet, H. (2006) Foundations of Multidimensional and Metric Data Structures. Morgan Kaufmann. ISBN 0-12-369446-9.

Ubick, D., Paquin, P., Cushing, P.E., and Roth, V. (eds) (2009), Spiders of North America: An Identification Manual. American Arachnological Society.

Vollrath, F., and Knight, D. P. (2001). "Liquid crystalline spinning of spider silk". *Nature,* 410(6828): 541–548.

Woodward Jr, J.D., Orlans, N.M., Higgins, P.T. (2003) Biometrics, McGraw-Hill/Osborne, 2003 – 432.

Yan Ma, Bojan Cukic, and Harshinder Singh (2004). A Classification Approach to Multimodal Biometric Score Fusion. Lane Department of Computer Science and Electrical Engineering, Department of Statistics West Virginia University.

Key Terms and Definitions

Biodiversity: is the diversity of all living forms at different levels of complexity: genes, species, ecosystems and even landscapes and seascapes.

Biometric analysis: refers to the analysis of the distinctive, measurable characteristics used to label and describe individuals.

Spider Cobwebs: is a device created by a spider out of proteinaceous spider silk extruded from its spinnerets.

Artificial Intelligence: is the study and design of intelligent agents.

Machine Learning: is the branch of the artificial intelligence that builds and study systems in order to carry out a defined task based on the experience.

Patter Recognition: in machine learning, is the process where labels are assigned to a given value that belong to a certain class.

Feature Extractor: is the method applied to obtain the discriminative information of the source analyzed.

References

Ahmed, N., Natarajan, T., and Rao, K.R. (1974). Discrete Cosine Transform", *IEEE transactions on Computes,* 90-93.

Akansu, A.N., Haddad, R. A. (1992). Multiresolution signal decomposition: transforms, subbands, and wavelets. Boston, MA: Academic Press.

Arlot, S. (2010). A survey of cross-validation procedures for model selection. *Statistics Surveys*, 4, 40–79.

Cardoso, P., Arnedo, M.A., Triantis, K.A., and Borges, P.A.V. (2010). Drivers of diversity in Macaronesian spiders and the role of species extinctions. *Journal of Biogeography,* 37, 1034-1046.

Carvalho, J. C., Cardoso, P., Crespo, L.C., Henriques, S., Carvalho, R., and Gomes, P. (2011). Biogeographic patterns of spiders in coastal dunes along a gradient of mediterraneity. *Biodiversity and conservation,* 1-22.

Comon, P. (1994). Independent Component Analysis: a new concept? *Signal Processing,* 36(3),287–314.

Eberhard, W.G. (1990). Early Stages of Orb Construction by Philoponella Vicina, Leucauge Mariana, and Nephila Clavipes (Araneae, Uloboridae and Tetragnathidae), and Their Phylogenetic Implications. Journal of Arachnology, *American Arachnological Society,* 18(2), 205-234.

Eberhard, W.G. (Sep., 1982). Behavioral Characters for the Higher Classification of Orb-Weaving Spiders, Evolution, *Society for the Study of Evolution,* 36 (5), 1067-1095.

Eberhard,W.G. (1969). Computer Simulation of Orb-Web Construction. *Journal of American Zoologist,* 9(1), 229-238.

Finch, O. D., Blick, T., and Schuldt, A. (2008). Macroecological patterns of spider species richness across Europe. *Biodiversity and Conservation,* 17, 2849-2868.

Haibo Yao, and Lei Tian. (2003). A genetic-algorithm-based selective principal component analysis (GA-SPCA) method for high-dimensional data feature extraction. Geoscience and Remote Sensing, *IEEE Transactions,* 41(6), 1469- 1478.

Jing, H., Si, J., Olson, B.P., and Jiping, H. (2005). Feature detection in motor cortical spikes by principal component analysis. Neural Systems and Rehabilitation Engineering, *IEEE Transactions,* 13(3), 256-262.

Johnston, J. M. (2000). The contribution of microarthropods to aboveground food webs: A review and model of belowground transfer in a coniferous forest. *American Midland Naturalist,* 143, 226-238.

Kevman, V. (2001). Learning and Soft Computing: Support Vector Machines, Neural Networks, and Fuzzy Logic models. Puiblished by The MIT Press.

Kuncheva, L.I. (2004). Combining Pattern Classifiers, Methods and Algorithms, Published by John Wiley and Sons.

Langley, P., Bowers, E.J., Murray, A. (2010). Principal Component Analysis as a Tool for Analyzing Beat-to-Beat Changes in ECG Features: Application to ECG-Derived Respiration. Biomedical Engineering, *IEEE Transactions,* 57(4), 821-829.

Lim, J.S. (1990). Two-Dimensional Signal and Image Processing, Englewood Cliffs, NJ, Prentice Hall.

Ma, Y., Cukic, B., Singh, H. (2004). A Classification Approach to Multimodal Biometric Score Fusion. Lane Department of Computer Science and Electrical Engineering, Department of Statistics West Virginia University.

Mallat, S. (1989). A theory for multiresolution signal decomposition: the wavelet representation. *IEEE Pattern Analysis and Machine Intelligence,* 11(7), 674-693.

Nan Liu, and Han Wang. (2006). Feature Extraction with Genetic Algorithms Based Nonlinear Principal Component Analysis for Face Recognition. Pattern Recognition, *18th International Conference,* 3, 461-464.

Otsu, N. (1979). A threshold selection method from gray-level histograms. *IEEE Trans. Sys., Man., Cyber.,* 9 (1), 62-66.

Pearson, K. (1901). On Lines and Planes of Closest Fit to Systems of Points in Space" (PDF). *Philosophical Magazine,* 2 (11), 559–572.

Peterson, A. T., Osborne, D. R., and Taylor, D. H. (1989). Tree trunk arthropod faunas as food resources for birds. *Ohio Journal of Science,* 89(1), 23-25.

Qingfu Zhang, and Yiu Wing Leung. (2000). A class of learning algorithms for principal component analysis and minor component analysis. Neural Networks, *IEEE Transactions,* 11(1), 200-204.

Schölkopf, B., and Smola, A.J. (2002). Learning with Kernels. Support Vector Machines, Regularization, Optimization, and Beyond. Published by The MIT Press.

Suresh, P. B., and Zschokke, S. (2000). A computerised method to observe spider web building behaviour in a semi-natural light environment. *19th European colloquium of arachnology,* 117-122.

Suykens, J.A.K., Van Gestel, T., De Brabanter, J., De Moor, B., and Vandewalle, J. (2002). Least Squares Support Vector Machines. Puiblished by World Scientific.

Sytnik, K.M. (2010), Preservation of biological diversity: *Top-priority tasks of society and state Ukrainian Journal of Physical Optics,* 11 (SUPPL. 1), pp. S2-S10.

Travieso, C.M., Botella, P., Alonso, J.B., and Ferrer M.A. (2009). Discriminative common vector for face identification. 43rd Annual 2009 *International Carnahan Conference on Security Technology,* 134-138.

Unesco.(2012).http://www.unesco.org/new/en/natural-sciences/special-themes/biodiversity-initiative.

Vapnik, V. (1995). The Nature of Statistical learning Theory. *Springer Verlag.*

In: Spiders
Editor: Marq Santerre

ISBN: 978-1-62618-517-3

Chapter 5

ANT MIMICRY IN SPIDERS – SHORT COMMUNICATION

Fadia Sara Ceccarelli*

Department of Environmental Science, Biogeography Section, University of Basel, Basel, Switzerland

ABSTRACT

Mimicry is a natural phenomenon which never ceases to amaze, involving one organism taking on characteristics of another, to gain evolutionary advantage. These strategies based on "deceit" carry with them advantages to the mimic (or deceiver) either in the form of heightened protection from potential predators (as in the case of Batesian or Müllerian mimicry), or easier access to prey (aggressive mimicry). Mimics exploit different sensory channels, including visual, acoustic or chemical systems, by looking, sounding or "smelling" like a model organism, respectively. Certain characteristics of model organisms are favorable to mimics, especially when the mimicry is a form of protection from predators. Among insects, an ideal group of model organisms are the ants, as they are numerous, often aggressive and unpalatable for many other animals. Ant mimicry is therefore common in several invertebrate families. In spiders, ant mimicry occurs in at least 13 different families. Here, ant mimicry in spiders will be outlined, briefly describing the main findings to date of this vast and fascinating topic.

* Corresponding author: Fadia Sara Ceccarelli. E-mail: saracecca@hotmail.com.

INTRODUCTION

The phenomenon of mimicry has long caught the attention of biologists, with the first documented mimetic organisms appearing in works of 19th century naturalists such as Darwin [1] and Bates [2]. Mimicry is broadly defined as the resemblance of one organism to another, taxonomically unrelated one, and carries with it some kind of "deceit". Mimicry involves a mimic (the organism carrying out the "deceit"), a model (on which the "deceiver" bases its "deceit"), and a target (or the "deceived"), which can be a group of organisms, sometimes including the model itself. The parties involved, and the nature of the deceit is what dictates the name given to the mimicry. For example, Batesian mimicry occurs when a palatable mimic gains protection from predators by mimicking an unpalatable (noxious or dangerous) organism, while in Müllerian mimicry there is no clear-cut palatable-mimic-noxious-model scenario, but rather a group of organisms reinforcing their unpalatability through resembling each other. When the target of the mimicry becomes the prey itself, we get what is known as aggressive mimicry, whereas Wasmannian mimicry facilitates the mimic living in close association to the model [3]. The mimic can exploit different sensory channels of the target, such as visual, acoustic or chemical, among others. Generally, the premises contributing to the success or stabilisation of mimicry in a population are that (i) the mimic must live in proximity to the model and (ii) the mimic should not be more abundant than the model.

Mimicry occurs in a large number of taxa, ranging from plants to vertebrates [4]. Due to their relatively short generation times and incredible abundance, arthropods display perhaps the most striking and varied examples of mimicry. Among terrestrial invertebrates, common model organisms are ant species. Ants are considered ideal model organisms due to important attributes such as their abundance, ubiquity, eusociality and unpalatability for many predators caused by their aggressive nature, bites and/or stings and in certain cases production of formic acid [3, 5, 6]. Ant colonies all over the world are therefore exploited to varying degrees by terrestrial invertebrates including other insects such as beetles (Coleoptera), bugs (Hemiptera), butterflies and moths (Lepidoptera). Apart from other insects, spiders have also repeatedly evolved associations of mimicry to ants. The ant mimicry found in spiders ranges from Batesian to Wasmannian and/or aggressive, and is mostly visual and chemical [5, 7, 8]. Proximity to the model – one of the prerequisites for mimicry to be effective – can be dangerous for the spiders, as they can become victims of the ants' aggression.

This is why most ant mimicking spiders tend to avoid contact with ants, unless their strategy is to enter the ants' colonies and live among their models. In this case the spider is referred to as a myrmecophile, and in some cases the mimicry facilitates preying on ants or ant larvae. In order to enter the ants' nests or live inside the colony undetected, the most effective strategy is for the spiders to "fool" the ants chemically (as in the case of *Cosmophasis bitaeniata* [9] and *Gamasomorpha maschwitzi* [10]). That is to say, the sensory channel exploited by the mimics is the model ants' chemosensory system, meaning that the mimicry is mainly focused on the cuticular hydrocarbons of ants.

Myrmecomorphs are to human observers perhaps the most striking of all ant-mimics. Myrmecomorphy refers to the physical resemblance to an ant, and is therefore mimicry that targets the visual system of other animals [5, 7]. The physical resemblance to ants is achieved through a combination of morphological and/or behavioural adaptations, some of which yield an astonishingly close resemblance to ants, while others are more of an approximation. The postulated explanations for this range of resemblance include (i) a subjective classification of "good" and "poor" mimics based on human observers, which may change depending on the predator taxon [11, 12]; (ii) a cost-benefit model whereby "approximate" resemblance to a model may suffice under circumstances related to the model [e.g. 13, 14, 15] or (iii) the various taxa of myrmecomorphs are at differing stages of predator-mediated selection, where some have had more time to adapt than others [e.g. 16, 17].

Ample evidence points to the fact that myrmecomorphic spiders gain protection from visual predators, thus being cases of Batesian mimicry [e.g. 18, 19, 20, 21], since a considerable number of vertebrate as well as invertebrate predators of spiders will avoid ants. In some cases, aversion to ants has even been shown to be innate [22]. Myrmecomoprhs obtain their ant-like appearance through a combination of the following morphological adaptations. A general body plan of three (instead of two) main segments achieved through the constriction of the carapace or abdomen; colorations and textures mimicking those of the ants, including sometimes colorations adding to the illusion of a three-segmented body plan; colorations of the carapace resembling the large, compound eyes of ants (eye spots); generally more elongated and slender body [5]. Some spiders also join the enlarged front femurs in front of their head, to recreate the illusion of a third body part [23]. This is an example of a part morphological, part behavioural modification. Several other behavioural modifications function to re-enforce the ant-like appearance. For example, myrmecomorphs may acquire an ant-like gait, and/or lift their first pair of legs and wave them around, recreating the illusion of ants' antennae [24, 25]. Some ant-mimicking spiders such as the jumping spider *Myrmarachne lupata* (Araneae: Salticidae) even modify their prey capture technique not to lose their similarity to ants, lunging on their prey, rather than jumping, as other salticids would do [26].

Batesian myrmecomorph ant-mimicking spiders are remarkable when it comes to strategies maximizing their evolutionary success. For example, to keep mimic numbers low, some species adopt polymorphism, where members of the same species mimic different model ant species through variations in colour [27]. Also, as myrmecomorphic spiders pass though their various instar changes, they sequentially mimic different ant species (a strategy termed transformational mimicry [28]) as a way of always being approximately the same size as the model. Usually, salticids are solitary, yet some species have been found in large aggregations, bringing forth the theory of "collective mimicry", or mimicking a whole ant colony at once [29, 30]. A morphological character that may appear to lower the resemblance of certain spiders to ants is the enlarged chelicerae of males such as those of *Myrmarachne smaragdina* (Figure 1; thought to be a sexually selected character), yet even in those cases, the spider may still be protected by resembling an ant carrying a parcel in its mandibles (termed "compound mimicry") [31].

A short overview of the most striking ant-mimicking strategies in spiders has been provided here. However, considering that ant mimicry can be found in at least 13 families of spiders with more than 15 instances of independent evolution, and considering the vast and varied strategies, this is a topic that draws a great deal of interest and deserves much attention.

Figure 1. Male *Myrmarachne rubra* with enlarged chelicerae.

After all, ant mimicry in spiders provides a perfect platform on which to carry out theoretical and empirical studies to answer fundamental questions of beahviour, ecology and evolution.

REFERENCES

[1] Darwin, C. (1859) *On the origin of species*. John Murray, London.

[2] Bates, H. W. 1862. Contributions to an insect fauna of the Amazon valley. Lepidoptera: Heliconidae. *Trans. Linn. Soc. London* 23: 495-566.

[3] Rettenmeyer, C. W. (1970) Insect mimicry. *Annu. Rev. Entomol.* 15: 43-74.

[4] Wickler, W. (1968) *Mimicry in Plants and Animals*. McGraw-Hill, New York.

[5] McIver, J. D. and Stonedahl, G. (1993) Myrmecomorphy: Morphological and behavioural mimicry of ants. *Annu. Rev. Entomol.* 38: 351-379.

[6] Hölldobler, B. (1971) Communication between ants and their guests. *Sci. Am.* 224: 86-93.

[7] Cushing, P. E. (1997) Myrmecomorphy and myrmecophily in spiders: A review. *Florida Entomol.* 80: 165-193.

[8] Cushing, P. E. (2012) Spider-Ant Associations: An Updated Review of Myrmecomorphy, Myrmecophily, and Myrmecophagy in Spiders. *Psyche* 2012: 1-23.

[9] Elgar, M. A. and Allan, R. A. (2006) Chemical mimicry of the ant *Oecophylla smaragdina* by the myrmecophilous spider *Cosmophasis bitaeniata* is it colony-specific? *J. Ethol.* 24 (3): 239–246.

[10] Witte, V., Leingärtner, A., Sabab, L., Hashim, R., and Foitzik, S. (2008) Symbiont microcosm in an ant society and the diversity of interspecific interactions *Animal Behaviour* 76: 1477–1486.

[11] Dittrich, W., Gilbert, F., Green, P., Mcgregor, P., and Grewcock, D. (1993) Imperfect mimicry: a pigeon's perspective. *Proc. Royal Soc. B* 251: 195–200.

[12] Théry, M. and Gomez, D. (2010) Insect Colours and Visual Appearance in the Eyes of Their Predators. *Advances in Insect Physiology* 38: 267-353.

[13] Gilbert, F. (2005) The evolution of imperfect mimicry. In: *Insect Evolutionary Ecology*, M. D. E. Fellowes, G. J. Holloway and J. Rolff, Eds., pp. 231–288, CABI, Wallingford, Wash, US.

[14] Kikuchi, D. W. and Pfennig, D. W. (2010) High-model abundance may permit the gradual evolution of Batesian mimicry: an experimental test. *Proc. Royal Soc. B* 277: 1041–1048.

[15] Speed, M. P. and Ruxton, G. D. (2010) Imperfect Batesian mimicry and the conspicuousness costs of mimetic resemblance *Am. Nat.* 176: E1–E14.

[16] Edmunds, M. (2000) Why are there good and poor mimics? *Biol. J. Linn. Soc.* 70 (3):459–466.

[17] Mappes, J. and Alatalo, R. V. (1997) Batesian mimicry and signal accuracy. *Evolution* 51: 2050-2053.

[18] Cutler, B. (1991) Reduced predation on the antlike jumping spider *Synageles occidentalis* (Araneae: Salticidae) *J. Insect Beh.* 4 (3): 401–407.

[19] Durkee, C. A., Weiss, M. R. and Uma, D. B. (2011) Ant mimicry lessens predation on a North American jumping spider by larger salticid spiders *Environ. Entomol.* 40(5): 1223–1231.

[20] Edmunds, M. (1993) Does mimicry of ants reduce predation by wasps on salticid spiders? *Mem. Q. Mus.* 33: 507-512.

[21] Nelson, X. J. (2012) A predator's perspective of the accuracy of ant mimicry in spiders *Psyche* 2012: 5 p. Article ID 168549.

[22] Nelson, X. J., Jackson, R. R., Li, D., Barrion, A. T., and Edwards, G. B. (2006) Innate aversion to ants (Hymenoptera: Formicidae) and ant mimics: experimental findings from mantises (Mantodea) *Biol. J. Linn. Soc.* 88(1): 23–32.

[23] Deeleman-Reinhold, C. L. (1993) A new spider genus from Thailand with a unique ant mimicking device *The Nat. Hist. Bull. Siam Soc.* 40(2): 167–184.

[24] Ceccarelli, F. S. (2008) Behavioral mimicry in *Myrmarachne* species (Araneae, Salticidae) from North Queensland, Australia. *J. Arachn.* 36 (2): 344–351.

[25] Reiskind, J. (1977) Ant-mimicry in Panamanian Clubionid and Salticid spiders (Araneae: Clubionidae, Salticidae). *Biotropica* 9: 1-8.

[26] Jackson, R. R. (1986) The biology of ant-like jumping spiders (Araneae, Salticidae) - Prey and predatory behavior of *Myrmarachne* with particular attention to *Myrmarachne lupata* from Queensland. *Zool. J. Linn. Soc.* 88: 179-190.

[27] Nelson, X. J. (2010) Polymorphism in an ant mimicking jumping spider. *J. Arachnology* 38(1): 139–141.

[28] Mathew, A. P. (1935) Transformational deceptive resemblance as seen in the life history of a plant bug (*Riptortus pedestris*), and of a mantis (*Evantissa pulchra*). *J. Bombay Nat. Hist. Soc.* 37: 803-813.

[29] Jackson, R. R., Nelson, X. J. and Salm, K. (2008) The natural history of *Myrmarachne melanotarsa*, a social ant-mimicking jumping spider. *New Zealand J. Zool.* 35(3): 225–235.

[30] Nelson, X. J. and Jackson, R. R. (2009) Collective Batesian mimicry of ant groups by aggregating spiders *Animal Beh.* 78: 123–129.

[31] Nelson, X. J., Li, D. and Jackson, R. R. (2006) Out of the frying pan and into the fire: a novel trade-off for Batesian mimics *Ethology* 112(3): 270–277.

Index

C

D

H

I

J

L

M

N

O

P

Q

R

S

T

U

V

W

X

Y